TESTING AND EVALUATION OF AGRICULTURAL MACHINERY

Testing and Evaluation of AGRICULTURAL MACHINERY

M.L. MEHTA
Northern Region Farm Machinery Training & Testing Institute
Hissar (Haryana)

S.R. VERMA
Department of Farm Power & Machinery
Punjab Agricultural University, Ludhiana

S.K. MISRA
North Eastern Farm Machinery Training & Testing Institute
Vishwanath Cherailli, Assam

V.K. SHARMA
Department of Farm Power & Machinery
Punjab Agricultural University, Ludhiana

2016
Daya Publishing House®
A Division of
Astral International (P) Ltd
New Delhi-110 002

Published by   :   **Daya Publishing House®**
A Division of
Astral International Pvt. Ltd.
– ISO 9001:2008 Certified Company –
4760-61/23, Ansari Road, Darya Ganj
New Delhi-110 002
Ph. 011-43549197, 23278134
E-mail: info@astralint.com
Website: www.astralint.com

PREFACE

Agricultural Mechanization is a sine qua non to remove drudgery, improve working comfort, enhance timeliness, reduce losses and increase production and productivity. Accordingly, use of better power viz. tractors and different types of agricultural machines in Indian agriculture has risen sharply on Indian farms to boost food and fibre production. But to safe guard the user's interest, to ensure better quality and reliability of machines and for sustained growth of farm machinery industry, there is a need for sound scientific testing and evaluation of farm machines by using instrumentation and accepted methodology. Thus, testing and evaluation holds the proper key to standardization and quality control of agricultural machinery for better acceptability and sustained farm production. To satisfy the genuine need of different sectors, this book has been prepared. It is expected to serve as a Text-Book for the students of Agricultural Engineering degree and Postgraduate degree programme. It may also serve the needs of professional engineers, scientists, testing institutions and Research Organizations dealing with testing and evaluation of agricultural machinery. This book will also cater to the needs of tractor and agricultural implement manufacturing industries, consultants, Agricultural Universities/Colleges as a valuable reference for quality improvement and standardization. It is hoped this book will be a valuable reference for different groups in developing countries of Asia and Africa and Latin America.

The book has been organised in twelve Chapters. The first and second chapters deal with the status of agricultural mechanization in India and Testing and Evaluation System in India respectively. Chapter-III deals with Testing and Evaluation of Tractors. Chapter-IV to X deal with testing and evaluation of different agricultural implement, irrigation pumps and plant protection equipment. Chapter-XI deals with data acquisition, processing and analysis by Personal Computers (PCS). Chapter-XII deals with Testing of Agricultural Machinery for safety.

The authors wish to express their gratitude to different organizations and individuals who have contributed directly or indirectly in the preparation of the manuscript of this book. Our sincere thanks are due to Dr C.P.Singh, Dr S.K.Sondhi, Dr I.K.Garg, Dr S.S. Ahuja, Dr D.S. Taneja, Prof. N.S.Sandher and Prof. P.K.Gupta. Assistance of BRAIN-JAPAN, Bureau of Indian Standards, NRFMT&TI, Hissar, NEFMT&TI, Vishwanath Cherailli, P.A.U., Ludhiana, RNAM, CIAE, CFMT&TI, Budni, FAO and Ministry of Agriculture is also sincerely acknowledged. We also thank National Agricultural

Technology Information Centre, the publisher for timely printing of this book.

We are conscious of the time constraint each one of us had in writing this book. It is a maiden effort of its kind. It is possible that the method of presentation of the information may be at variance from that of some other authors. But what we thought best, we adopted. We hope, the readers will find this book informative and useful. The authors will no doubt appreciate the constructive criticism and suggestions from the readers.

M.L. Mehta
S.R. Verma
S.K. Misra
V.K. Sharma

August, 1995

TABLE OF CONTENTS

PREFACE

1 Status of Agricultural Mechanization in India   1

Introduction, Level of Mechanization in India, Agricultural Machinery
Situation, Mechanization policy, Future Prospectus

2 Testing and Evaluation System in India   12

Intrdouction, Standardization efforts, Testing Programme and
Procedure, Type of testing systems

3 Testing and Evaluation of Agricultural Tractor   18

Introduction, General regulations, Terminology, Basic Measurements,
Test items, Testing Procedure

4 Testing and Evaluation of Tillage Machinery   56

Introduction, Type of Tests, General Conditions, Testing Procedure,
Enterpretation of test results

5 Testing and Evaluation of Seed-cum-Fertilizer Drills   69

Introduction, Functional requirements of Seed-cum-Fertilizer Drills,
Major components, Critical components, Type of tests, General
conditions, Test conditions, Procedure for laboratory testing,
Procedure for field testing

6 Testing and Evaluation of Rice Transplanter   80

Introduction, Classification of rice transplnter, Nursery for transplanter,
Construction of rice transplanter, Scope of test, Terminology, General
conditions, Test conditions, Test items, Some suggested limits of rice
transplanter

7 Testing and Evaluation of Irrigation Pumps   89

Introduction, Construction of Centrifugal Pump, Terminology,
Selection of Parameters, Causes for low overall efficiency of pumping
system, Data required for selection of centrifugal pump, Quality criteria,
Selection of pumpset, Guarantees and purpose of testing, Testing
procedure, Testing parameters, Measuring instruments , Calculation of

viii

performance parameters, Case study for analysis of testing data,
Verification of guaranteed duty points

**8 Testing and Evaluation of Manually Operated Sprayers and   110
    Dusters**

Introduction, Type of pesticide application equipments, Definition of
important terms, Type of tests for sprayers, Testing method for sprayers,
Type of tests for Dusters, Testing method for dusters

9 Testing and Evaluation of Combine Harvester   128

Types of grain combines, Combine systems, Test items, Procedure
for laboratory testing, Procedure for field testing, Important field terms,
Material required for measurement and sampling during field test,
Observations to be recorded during and after test, Sample analysis,
Data analysis, Summary of performance parameters required for
analysis during field test

10 Testing and Evaluation of Power Thresher   151

Introduction, History of development of threshing devices, Principles
of operation of a threshers, Terminology, Type of tests, Test
Procedure, Performance requirement, Computation of results,
Determination of corrected output capacity, Determination of power
consumption, Checking for safety

**11 Data acquisition, Processing and Analysing by Personal   166
    Computers**

Introduction, Data acquisition, Measuring devices, Data analysis,
General approach in computer problem solving, Computer
components, Software, Operating system Writing of programmes

12 Safety Testing of Agricultural Machinery   179

Introduction, Type and causes of agricultural machinery accidents,
Technical requirements for ensuring machinery safety, Testing of
Agricultural Machinery for safety, Checking of tools and devices,
Method of safety testing, Definition of terms, Test Procedure,
Criteria for acceptance of ROPS, Safety precautions

Appendices   206

Bibliography   247

Chapter 1

STATUS OF AGRICULTURAL MECHANIZATION IN INDIA

1.1 INTRODUCTION

Most of the developing countries of Asia have the problem of high population and low level of land productivity as compared to the developed nations. One of the main reasons for low productivity is insufficient power availability on the farms and low level of farm mechanization. This is especially true for India.

It is now realized the world over that in order to meet the food requirements of the growing population and rapid industrialization, modernization of agriculture is inescapable. The package of modern technology, besides, improved seeds and fertilizers, includes use of efficient and economical farm implements, machines and suitable form of farm power. It is said that on many farms, production suffers because of improper seedbed preparation and delayed sowing, harvesting and threshing. Mechanization enables the conservation of inputs through precision metering ensuring better distribution, reducing quantity needed for better response and prevention of losses or wastage of inputs applied. Mechanization reduces unit cost of production through higher productivity and input conservation. Above all, mechanization removes drudgery/hazards to human being and animal and provides respectability and dignity to labour.

Agricultural implement and machinery programme of the Government has been one of selective mechanization with a view to optimize the use of human, animal and other sources of power. With the increase in the area under irrigation with multi-cropping/Intensive cropping, requirement of farm power and agricultural implements has considerably increased. In order to meet the requirements, steps were taken to increase availability of implements, irrigation pumps, tractors, power tillers, combine harvesters and other power operated machines and also to increase the production and availability of improved animal drawn implements. Special emphasis was laid on the later as more than 70% of the farmers fall in small and marginal category. However, no definite policy for agriculture mechanization has emerged either by defining selective mechanization or by fixing mile stones. Adhoc programmes with limited targets were taken up/continued and discontinued at convenience. On the other hand, industry has taken initiatives to manufacture new items and marketed them. Liberal credit has also helped in acquiring new machines. However, this has led to uneven mechanization to some extent. For example, Faridkot district in Punjab recorded 137 tractors per thousand hectare in 1986-87 whereas many of the districts in the country may not have a single tractor even today. The availability of farm power through mechanical means, was estimated as 2.71 hp per hectare in Punjab in 1986-87 whereas many states may not have even one tenth of it.

It is generally said that mechanization of small farms is difficult. But Japan having

Table 1 : Density of tractor and combine harvestors in selected countries

Sr. No.	Country	Tractor population (Number per 1000 ha)	Combine Harvester/population (Number per 1000 ha)
1.	India	5.59	0.01
2.	China	9.27	0.39
3.	Japan	403.76	303.14
4.	Pakistan	9.02	0.04
5.	Germany (FR)	195.77	18.83
6.	U.K.	77.33	8.07
7.	USA	24.85	3.40
8.	Sri lanka	15.10	-

Note : Data given in Table 1 does not take into account the average size of the machine

Source : Verma & Sharma, 1991

Table 2 : Level of mechanization in India

Machinery centage of	Status of mechanization (estimated)		
	Population (lakhs)	Average command area (ha)	Per- total net area
Tillage			
Tractor	10.7	15	11.80
Iron plough (animal)	10.5	2-3	15.44
Seed Drill			
Tractor	5.0	15	5.51
Animal	147.0	2-3	32.43
Irrigation Pumpsets	109.0	4-5	30.08
Threshers			
Wheat	18.0	5-6	44.12
Paddy	1.5	6-8	3.31
Multicrop	1.6	5-6	1.10
Harvesting			
Reaper	0.2	15	0.61
Combine	0.475	100	0.37
Plant protection equipment	26.5	12	23.38

Source : Gill, 1993

average land holding even smaller than ours, with proper mechanization has led agriculture to great heights. For small holdings, small machines are available at reasonable prices in other parts of the world. In order to minimize the drudgery of small farmers, to increase efficiency and save farmer's time for taking up additional/ supplementary generating activities, the use of modern time saving machines/ implements of appropriate size need to be suitably promoted.

1.2 Level of mechanization in India

Industrialized countries of the West and in the Asian sub-continent have achieved almost 100% mechanization in agriculture. Among the developing countries even China, South Korea and Pakistan are much ahead of India. The density of machines for seed bed preparation and for harvesting could be the indicators for comparing the levels of mechanization in different countries. Facts recorded in from FAO Year Book 1990, indicating the density of tractors and combine harvestors per thousand hectare may be seen at Table 1. The world average is 19.15 tractors per thousand hectare and the highest is 403.76 tractors per thousand hectare in Japan. Even in Asia as a whole, there are 12.72 tractors per thousand hectare. The tractor density in India is only 5.59 per thousand hectare, which is far below the International average and even below the neighbouring countries like China, Pakistan and Srilanka.

The world average density of combine harvester is 2.87 units per thousand hectare whereas Japan has the highest of 303.I4 units per thousand hectare. In India, this stands at 0.01 units per thousand hectare which is lower compared to China and Pakistan. Effect of power input on yield reported for different countries is shown in Fig.1.1.

In India, the introduction of improved implements was initiated in 1880, with the advent of the Department of Agriculture. Regular demonstrations were conducted on the farmer's fields in the then Presidency of Bombay, Madras and Calcutta. During early 20th century, efforts continued mostly on animal drawn ploughs, harrows, ridger, cultivators, ridger plough and scoops through imports. Indian manufacturers namely M/s Kirloksar Brothers, M/s Burn & Co. and M/s Cooper Engineering Works, took the lead for import and manufacturing of these implements. With the organization and expansion of the Departments in the year I905, steps were taken to accelerate the pace of introduction of improved farm implements. As regards to tractors, the first steam model was brought by Sir Joginder Singh in 1914, in his State in Lakhimpur Distt. of Utter Pradesh. Shri Khase Rao Jadav brought the first petrol engine tractor in 1920. In I923, I5-30 hp gasoline tractors were purchased by the then Central Province for Kans eradication. However, a modest beginning of mechanization was made by utilizing World War-II surplus machines by the Central Tractor Organization in late 40s and early 50s.

Manufacture of tractors and power tillers in India commenced in 60s whereas

4

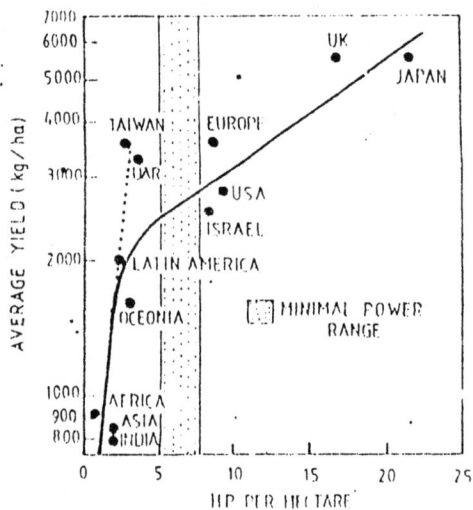

FIG.1.1 RELATIONSHIP BETWEEN YIELDS AND
POWER OF MAJOR FOOD CROPS

Source: Anon.,1978

manufacture of engines and pumps started much earlier. Till mid 70s, part of the demand of farm machines was met through imports. Subsequently, most of ag ricultural machinery demand is being met from indigenous sources (Table 2). As a result of sustained efforts, sale of tractors and power tillers during 1991-92 stands at 1,50,000 and 7,500 respectively.

Though substantial units of agricultural machines have been introduced in the recent past, yet adoption had mostly been in the northern States and in the scattered pockets/areas where better irrigation facilities were available. Further, to understand the subject in a proper perspective, particularly in terms of the net area covered with the use of machinery, the level of mechanization of the major farm operations is shown in Table 2.

1.3 Agricultural Machinery Situation

A. Production System
i) Tractors

In India, tractors are manufactured in the organized sector. There are 19 units manufacturing tractors of which seven contribute for about 90% of the annual production. The production of tractor s started in the year 1961-62 when 880 numbers were produced. It had gone to 151,759 numbers in 1991-92 as shown in Table 3.

Table 3. Production and sale of tractors.

Year	Production	Sale
1987-88	92930	93157
1988-89	109987	110323
1989-90	121624	122098
1990-91	139233	139831
1991-92	151759	150582

Source : Gill, 1993

The pattern of the sales shows that the State of U.P., Punjab and Haryana account for more than 50%, of the total sale. The sale of tractors in the State of Rajasthan, Madhya Pradesh, Bihar, Maharashtra and Gujarat had increased considerably in the recent years.

ii) Power Tillers

There are two units in the organized sector and five in the small scale sector manufacturing power tillers. The production and sale of power tillers during the last five years is shown in Table 4.

Table 4. Production and sale of power tillers

Year	Production	Sale
1987-88	3005	3097
1988-89	4798	4678
1989-90	5334	5442
1990-91	6228	6316
1991-92	7580	7528

Source : Gill, 1993

iii) Combine Harvesters

Self-propelled combine harvesters for harvesting cereal crops are being manufactured by 9 units. The production and sale during the past five years is shown in Table 5. Tractor drawn combine harvesters and reapers are also being manufactured in the small scale sector.

iv) Implements

Manufacturing of implements is reserved for small scale sector and the estimate

Table 5. Production and sale of Combine Harvestor

Year	Production	Sale
1987-88	149	144
1988-89	93	94
1989-90	181	218
1990-91	352	363
1991-92	187	189

Source : Gill, 1993

annual turn-over is around Rs. 300 crores. The concentration of manufacturing units is in the States of Madhya Pradesh, Utter Pradesh, Punjab, Haryana, Bihar and Karnataka. In addition to about 15000 registered small scale units, there are over 2000 unregistered units in the country manufacturing handtools and small implements. Some of the State Agro Industries Corporations namely, Gujarat, U.P., M.P., Andhra Pradesh, Bihar and Maharashtra, have undertaken manufacturing of improved agricultural implements.

B. Delivery systems

In the case of tractors, power tillers and combine harvesters manufacturers have appointed dealers in different parts of the country to look after sales and services. Some of the State Agro Industries Corporations are also agents for sale and services of popular brands of tractors and power tillers. There is a net work of about 2150 dealers for tractors and 320 dealers for power tillers. In the case of hand tools and agricultural implements, the manufacturers themselves undertake sale, repair and after sale services.

C. Pricing system

There is no statutory price control for farm machinery. The prices are governed by market forces. However it has been observed that annual increase of prices of tractors and power tillers during the last 5 years have been in the order of 11% for tractors upto 50 hp, 15% for tractors above 50 hp and 7% for power tillers.

D. Training system

Farm machinery worth about Rs. 4000 crores are sold annually in the country. These include about 1,50,000 tractors, 7,500 power tillers, 200 combine harvesters, 700000 irrigation pump sets, 1,50,000 power threshers and large quantity of plant protection equipments, agricultural implements and hand tools of various types etc. The farmers ought to be advised not only on the availability of these machines but also towards their proper selection, economical usage, efficient operation,maintenance, energy conservation and safety aspects. All these require intensive on-the-job demonstration and training to the users/owner and extension agencies. Training of extension functionaries is presently arranged by the Directorate of Extension, (Govt. of India) through Extension Education Institutes, ICAR Institutes, Krishi Vigyan Kendra (KVKs), State Agricultural Universities, Farmers' Training Centres and Extension Training Centres. Some of the Agricultural Universities and ICAR Institutes are being developed as Centres of advance training. However schedule of training courses organized by these institutes needs revision time to time.

Government of India have set-up four Farm Machinery Training & Testing Institute, at Budni (MP), Hissar (Haryana) , Garladinne (AP) and Biswanath Chariali (Assam). With the addition of two new Institutes in Rajasthan and Tamil Nadu in near future, training capacity may rise to about 6000 from the present level of 2200 persons in a year. Indian Institute of Management, Ahmedabad has estimated requirement of training for first time buyers of machines and for refresher training at l.37 lakhs by 2000 A.D.

It has been observed that in the Asian Countries where agriculture is one of the dominating sectors of the economy, training on agricultural machines is arranged regularly. Agricultural Machinery Institutes in Republic of Korea, Sri Lanka, Thailand, Bangladesh. Pakistan, Iran. Philippine and Japan etc. organize intensive training programmes on farm machinery for farmers, technicians and others. Experiences of these countries show that such training programmes had contributed significantly to expand farm mechanization and consequently increase in production and productivity.

E. Research & Development system

The Indian Council of Agricultural Research (ICAR) is the main organization looking after all agricultural research, including agricultural implements and machinery. It coordinates a number of research projects with centres at different places in the country. Adhoc plans are also financed by I.C.A.R. Some of the State Governments have also facilitated in setting up of research organizations at state level. However, a major portion of research on agricultural machines and mechanization is being conducted by different agricultural universities in the country. Each of the major states has atleast one agricultural university. The extension work on farm machines and implements is being carried out by state governments, agricultural universities, Council of Scientific and Industrial Research and manufacturers of agricultural machinery. Some of these agencies have their own research wings or finance research project on farm machines and mechanization.

A research programme usually concentrates on the development of equipment suitable to a given farming conditions. Generally, adaptive research is undertaken instead of starting the design afresh.The objective is to improve upon the performance of indigenous implements or develop a new implements that can either enhance labour productivity or appropriately mechanize the operation where a labour or power shortage hinders completing the task in time.

Major tractor manufacturers have set-up their own R&D facilities with well equipped laboratory including test-track. The Escorts, Eicher, Mahindra & Mahindra, HMT, TAFE, PTL etc. have adequate R&D facilities. Some recent tractor models owe their development to local R&D effort, including some fuel efficient diesel engines and heavy duty hydraulic system. However, the small scale industries hardly have any facility for research and development. Most of the items being manufactured by them have been adopted from the designs available within and outside the country.

Mechanization towards Greater Employment and Income

Investigations and surveys have revealed that farm mechanization plays a positive role in increasing direct and indirect employment and earning potential in rural areas through generation of opportunities for skilled operators, mechanics, salesman etc. Table 6 shows the increase in the earning capacity of various categories of workers engaged in agricultural mechanization related activities as well as the traditional unskilled labourers.

It is obvious that the rural man-power employed in comparatively higher technology based equipment and machinery activity earn substantially higher than those working on traditional activity including animal operated farming.

I.4 Mechanization policy

The country's current level of food production is just about balancing the rising population, but if it does not keep pace with the latter in the coming years, it may result to a serious food shortage.

In many of the developed and developing countries, a clear National Policy on Farm Mechanization has been evolved and the programmes with targets have been clearly defined. For instance in Japan, a law for promotion of Agricultural Mechanization was enforced in 1966. Similarly, in the Republic of Korea, Agricultural Mechanization promotion Law was enacted and made effective in 1978. Systematic implementation of programmes on production, research, inspection, safety, training, dissemination of information and utilization of agricultural machinery have been initiated by both these countries and have developed time-bound programmes to achieve 100% farm mechanization for plains, slope and mountainous agriculture. In the case of the Republic of China, no specific law appears to have been promulgated so far but the programme have been developed in a way similar to that of Japan and Korea to promote efficient mechanization within a stipulated period.

It is often argued that Indian Agriculture is dominated by small farmers and the use

Table 6 : Increase in earning capacity of skilled and unskilled labourers in India

S. No.	Type of labour	Increase in earning capacity (%)
1	Irrigation pump/thresher operator vs unskilled agricultural worker	63.4
2	Way-side tractor workshop vs unskilled agricultural worker	19.7
3	Dealer workshop helper vs unskilled agricultural worker	40.8
4	Tractor operator vs ploughman	53.8
5	Tractor mechanic vs village artisan	59.0
6	Sales vs village artisan	45.4
7	Spare parts store keeper vs village artisans	40.9

Source : Mehta, 1986

of machines may not be desirable. One school of thought also speaks of utilization of animal power for all agricultural operations. However, it can be seen that though there are 76.4% of small and marginal holdings but they command 28.7% of the operational area. Whereas, large farms and the medium farms, though comparatively fewer in number, command about 3/4th of the total cultivated area. These farms have marketable surplus food to feed the non agricultural population and also to a greater extent marginal farmers. These holding require to be mechanized not only to sustain agriculture but to improve productivity and reduce the cost of production. Productivity level in Japan, South Korea and in China are far higher than those in India. Even in the case of Punjab, where average land holding is about 3.77 hectares, significant improvements have been reported due to modernization of agriculture especially mechanization of agriculture. Punjab is contributing presently about 60% of wheat and 45% of rice to the Central Pool. Punjab is also providing employment to migratory labour from other states.

Two distinct farm mechanization models are available from the industrialized world. The first is the Japanese model based on small individually-owned farm machines. This model has now been successfully introduced in the Republic of Korea and Taiwan. The Japanese mechanization model is characterized by small, low weight machines with focus on optimum utilization of available labour and other agricultural inputs. The Western model is characterized by large horse power machines, developed for use in area where dryland cultivation is practiced. This model is popular in North America, Europe and Australia. Western style of agricultural mechanization model is prevalent in those countries where population density is not high and the field conditions at tillage and harvesting time are firm to support large machines. In case of Japanese model, it is practiced on small farms and on wet (soft) lands that are available during planting and harvesting period.

The VIIIth Indian National Five Year Plan strategy for development of mechanization provides for (a) increasing replacement of traditional and inefficient implements and hand tools by improved animal drawn implements (b) owning tractors, power tillers, etc. individually (c) providing custom services for power machines and, (d) supporting services of training, testing and research etc. With this strategy and the programmes to be taken up during the VIIIth Five Year Plan period, the annual growth of mechanization is expected to increase by not more than 1%.

1.5 Future Prospects

Technology in the developed countries has undergone sea change in recent years. Products being manufactured in India require a similar approach to provide more reliable machines in terms of economy in operation, comfort, safety, easy maintenance and higher efficiency. For example, the internal combustion engine require replacement of conventional pump and nozzle system to a single micro chip control system with electronic injection. Injection of controlled fuel would emit lesser smoke and better fuel efficiency. Turbo charging and super charging of the engines have become quite common now a days in the developed countries. Similarly, synchromesh transmission system on agricultural machines have become a common feature. Fluid couplings or turbo clutches are being incorporated to cushion both engine and transmission against shock load, jerking, vibrations and reducing clutch wear. Monitoring and control systems are also needed on machines to assist the operator by way of automation in control and informations on wheel slip, area covered, maintenance requirement etc. These developments are required for tractors, power tillers, combine harvesters, engines and other similar machines.

Considering the scope for introduction of new machines in our agriculture and the needs assessed by the State Governments, the following new machines may be needed in future:

1. Low cost small tractor suitable for dryland and wetland cultivation.
2. High clearance tractor for cotton and sugarcane crops.
3. Self propelled paddy transplanter
4. Cotton picker.
5. Straw combine
6. Sugarcane equipment: planter with fertilizer attachment, cultivator for interculture, motorized circular saw for harvesting, top stripper and trash cleaner, bush cutter to trim cane stubbles for ratoon crop, set cutters and harvester.
7. Vegetable equipment: transplanter for Tomato, Chillies, onion, brinjal, cauliflower, etc., seed extractor, pneumatic seed-cum-fertilizer drill, grading and processing machines.
8. Horticulture equipment: Hydraulic platform, post hole digger, tree shaker.
9. Forestry equipment: pit digger, tree cutter, equipment for log handling and transplanting.

10. Forage machinery-self propelled forage harvester. reaper windrower, hay conditioner, hay rake.
11. Aero blast sprayer.
12. Potato combines
13. Sunflower combines

Indian Farm Machinery Industry has not made significant achievement in exports except a small quantity of tractors. Therefore, tractor and farm machinery manufacturers will have to strive for marketing in the world wide competition market to get reasonable market share in the exports.

Chapter 2

TESTING AND EVALUATION SYSTEMS

2.1 Introduction

Agricultural inputs such as improved seeds, chemical fertilizers and pesticides are no doubt essential for increasing crop yields, but equally important are the implements which ensure best possible use of these inputs and ensure their optimum doses for higher productivity of land. Without implements, the efficiency of application of the other inputs may get reduced and the overall productivity also declines. For instance, how so ever effective a pesticide may be, if it is not uniformly and adequately sprayed in the field with the help of a suitable sprayer, its effectiveness is lost. Even for operations such as land preparation, sowing and harvesting, machinery can make all the difference although the cost of some of these machines may be higher. However, cost reduction at the expense of quality will always be undesirable. The manufacturing cost is primarily a commercial problem and is linked up with a number of parameters.

Testing and evaluation are undertaken to quantify performance of machine for the desired operation. However, testing is defined as analysis of behaviour of machine when compared with standard codes/norms under ideal and repeatable conditions. On the other hand evaluation involves measurement of performance under actual field/ working conditions. Evaluation also encompasses the economic and social aspect in addition to the functional performance of the machine.

2.2 Standardization efforts

The International Organization for Standardization (ISO) is the apex body in the area of standardization at International level and has its membership on National Standards Bodies of various countries. The Bureau of Indian Standard, BIS (Formerly Indian Standard Institution) since its inception in 1947, realized the importance of standardization at National and International level. The Bureau has, therefore, from the beginning had been collaborating actively with ISO. In a developing economy like, in India, industrial development in general and agro/industrial in particular, makes its imperative to make an efficient use of all natural resources and organizational, technical and economic means. For this purpose, a coordinated, rational and efficient management in resource mobilization is of utmost importance. Experience has shown that standardization serves as an effective instrument in achieving this objective.

In the context of farm machinery, it has been observed that acceptance of farm machinery by the farmers largely depends on their quality. Hence, in order to reap the benefits of standardization including manufacture of high quality products, a need was felt for preparation of Indian Standards for agricultural machinery. Organized efforts in this direction were made by the Bureau of Indian Standards in late 50's by way of setting up a Technical Committee for formulation of standards for this group industry. The committee is generally consisted of representatives of Government departments, research, education and testing institutions and the manufacturing industries. So far, more than 300 standards have been published in this area. Since, in the past, manually operated and animal drawn implements were largely used and produced, to start with, such implements were selected for the purpose of standard-ization. Later on power operated equipment as well as tractors were taken up for standardization. Some of the highlights of the work are summarized below:

A. Standardization of specifications and interchangeability of components

When certain components of an implement or machine wear out, it remains under break-down for the period until the required component has been procured from the original manufacturer. This situation makes the repair, service and operation of the machine very difficult and uneconomical. The problem becomes more acute in case of imported items. Hence, the standardization of component was considered an important part of the work. Standards on components such as agricultural disc, reversible shovel, guards, knife sections, three-point linkage of agricultural tractors, power-take-off shaft of agricultural tractors, linch pin, ball and socket assembly, plough shares, shovels, sweeps, cultivator tines, spools for harrow, seed and fertilizer metering mechanism, chaff cutter blades, roller and axle for sugarcane crusher, nozzles, lances and cut-off device for sprayer, etc. have been brought out from the angle of interchangeability.

B. Implementation of Standards

All the efforts in formulation of standards would go waste if these are not implemented by the manufacturers, purchasers, leasing agencies and testing authorities. Implementation of standards would not only help in producing and procuring quality implements but would also reveal difficulties, if any, in adopting them and thus provide a feed-back to make the standard more realistic.

C. Certification Marks Scheme

To make quality implements and spares available to the farmers, Bureau of Indian Standard operates a Certification Marks Scheme, under which licenses are issued to such manufacturers who apply for use of ISI mark on their goods, to indicate that the quality is in conformity with the relevant Indian Standard. The BIS Certification Marks Scheme is operated on voluntary basis. Availability of certified agricultural machin-ery in the country could only be promoted by developing a demand for quality certified equipments through organized purchasers insisting on ISI Marked products.

Advantages of Certification Scheme

For Manufacturers
- Streamlining of production processes and introduction of quality control system.
- Independent audit of quality control system by BIS.
- Reaping of production economies accruing from standardization.
- Better image of products in the market, both internal and overseas.
- Winning for wholesaler's, retailers and stockists consumer confidence and goodwill.
- Preference for standard-marked products by organized purchasers, agencies of Central and State government, local bodies, public and private sector undertaking, etc. Some organized purchases offer even higher price for standard marked goods.
- Financial incentives offered by the Industrial Development Bank of India (IDBI) and Nationalized banks.

For consumers

- Conformity with Indian standards by an independent technical, national level organization.
- Helps in choosing a standard product.
- Free replacement of standard-marked products in case of their being found to be of substandard quality.
- Protection from exploitation and deception.
- Assurance of safety against hazards to life and property.

For Organized purchasers

- Convenient basis for concluding contracts.
- Elimination of the need for inspection and testing of goods purchased, saving time, labour and money.
- Free replacement of products with Standard Mark found to be substandard.

For Exporters

- Exemption from pre-shipment inspection, wherever admissible.
- Convenient basis for concluding export contracts.

For Export Inspection Authorities

- Elimination of the need for exhaustive inspection of consignments exported from the country, saving expenditure, time and labour.

2.3. Testing Programme and Procedures

In order to obtain accurate and repeatable results, it is most important to conduct the testing under a set of standardized test conditions and to follow standardized test procedures and formats. Attempts in this direction were made to develop detailed test codes since unorganized testing of machinery was posing problems for rationalizing the performance of various equipment. Test codes and procedures for agricultural tractor, mouldboard plough, disc harrow, seed-cum-fertilizer drill, power thresher, maize sheller, sugarcane crusher, seed cleaner and paddy weeder, chaff cutter, etc. have already been formulated and published by the Bureau of Indian Standards (BIS).

A. History of Tractor Testing

The history of agricultural tractor testing in the world is only 75 years old. The first tractor with an internal combustion engine was introduced in the American agriculture in the year 1889. The tractors were judged for their performance mostly in demonstrations and fairs on the basis of the comparative pulling power of the tractor for a certain number of bottoms of the mould board plough. The so called rating was termed as "Plough Rating". It was for the first time in year 1919 that an American farmer and senator of Nebraska State raised his voice for an Act for Compulsory and official testing of tractors to enable a check on unjustified claims of the tractor manufacturers. The tractor testing Act was passed in U.S.A. in 1920 to protect Nebraska farmers against such exploitations. The Agricultural Engineering Department in the Nebraska University was assigned the responsibility of establishing a Tractor Testing Laboratory and issuing the tractor test reports. Need for similar work in the interest of mechanized farming was felt simultaneously in other European countries. The National Institute of Agricultural Engineering or Testing Stations were established in important countries as under:

1. National Institute of Agricultural Engineering, U K
2. Swedish National Testing Institute of Agricultural Engineering, Sweden.
3. Agricultural Machinery Testing Station, Germany.
4. Centre National de Mechanique Agricole, Italy.
5. University of Saskatchewan, Canada.
6. Tractor Testing Station, School of Engineering, Victoria (Later shifted to Agricultural Engineering Department in the University of Melbourne), Australia.

By 1959, many more countries felt the need to test tractors either with an export interest or for finding out the utility of imported tractors under their special agro-climatic conditions.

The introduction of tractors in Indian agriculture started with imports in 40s in a small way. Imports gained a momentum in 50s when innumerable brands of tractors

16

were imported. Till 1960 India had to depend upon the imported tractors and the evaluation of the performance of the imported tractors was known mostly from their field performance in the Central Tractor Organization and from mechanical cultivation schemes of deferent States. In 1954, a need was recognized for having a National Tractor Testing Station in the country to test the suitability and performance of a model before large scale import and also to test tractors which.were proposed for indigenous manufacture. After temporary location at Nagpur, the Tractor Testing Station was finally established at Budni in 1959. Effective work on testing started in 1961 when the first tractor test report was released.

2.4 Type of Testing Systems

Different testing system currently invoke in India includes the following:

i) National Testing
Under this, the tests are conducted by the Farm Machinery training and Testing Institutes under the jurisdiction of Ministry of Agriculture, Government of India. Currently such institutes are located at Budni (MP),. Hissar (Haryana), Anantpur (A.P.)and Vishwanath Chariali (Assam) and are fully functioning. Type of tests conducted under the National Testing programme have been discussed in the subsequent paragraphs.

ii) Prototype Testing
Prototype testing includes the testing of research prototypes as well as the testing of production prototypes. Currently different State Agricultural Universities and ICAR Institutes with Engineering Divisions conduct prototype testing of their research equipment. Some of these institutions also undertake testing of production proto-types manufactured by different firms.

iii) Testing for quality marketing
This programme is carried out by the Bureau of Indian Standards under their Certification Marks Scheme through their Central Laboratory and other approved test laboratories located all over the country.

The following types of regular national tests are conducted on tractors, power tillers and agricultural machinery:

(a) Confidential test
Confidential test is meant for providing confidential information on performance of machines which are required for commercial production or to provide any special data that may ·be required by a manufacturer/applicant. The following categories of machines are covered under the scope of the confidential tests:
    i) a prototype model before it is ready for commercial production.
    ii) an improved model prior to it progressive manufacture/import on large scale.

iii) a machine under commercial production but with modification of one or more systems for improved performance.

iv) the machines submitted for test under BIS Certification Marks Scheme.

A Confidential test report can not be used for a commercial purpose. An applicant is not permitted to publish the confidential test report in full or in abbreviated form or to divulge the test results contained therein to any person or body.

(b) Commercial test

Commercial test is conducted on the machines which are ready for commercial production to establish their performance characteristics. The following types of commercial tests are undertaken:

i) Initial commercial Test

On indigenous or imported prototype machines ready for commercial production.

ii) Batch test

In view of the need for continuous improvement in the quality an indigenous machine under commercial production is tested after a certain time interval. At present this is applicable to tractors only. Every tractor model is required to be tested at Budni Institute once every two years through an executive order of the Government.

iii) Series test

Testing of different makes/models of similar type of machine simultaneously under identical conditions with a view to assess their comparative performance for the guidance of users, manufacturers and extension agencies.

iv) Test as per OECD Code

This is conducted on machines (which have already undergone initial commercial test) on specific request of the applicant/manufacturer exclusively for export purposes. Commercial test report can be published in full without any alteration or omission.

v) Users Survey programme

This forms an essential component of Batch Testing and designed for assessing the general performance, field complaints and durability of the machine in use with the farmers and other organizations. It also provides information on standard and efficiency of pre- and after sales-services being provided by the manufacturer and its dealers net-work.

Chapter 3

TESTING AND EVALUATION OF AGRICULTURAL TRACTORS

3.1 Introduction

Agricultural tractors are being increasingly used in India for mechanizing farm operations. Indeed the tractor industry has developed as one of the major engineering industries in the country. This has necessitated application of testing and evaluation of their performance on an uniform and rationalized basis. Agricultural tractor is a sort of self-propelled vehicle having atleast two axles with pneumatic or steel wheel or with tracks designed to carry out the following operations, primarily for agricultural purposes:

- To pull the trailers
- To carry, pull or propel agricultural tools or machinery and, where necessary, supply power to operate them with the tractor in motion or stationary.

3.2 General Regulations

1 Selection

The tractor submitted for commercial test (sample tractor) shall be taken at random from series production preferably by the representative of the testing institute, in consultation with the applicant and as per requirement of the test code.

The sample tractor shall be a production model in all respects, strictly conforming to the description and specification sheet submitted by the manufacturer/applicant. This description must define the model being tested.

The testing of a preproduction model tractor is permitted under special circumstances. In such case, the testing institute must certify in the test report that it has been checked that the series production confirms to the specification of the tested tractor.

The test report shall also state the mode of selection of tractor.

2. Manufacturer's Instructions

Once the test has started, the sample tractor shall never be operated in a way that is not in accordance with the manufacturer's published instructions which may be in the form of an operation manual, repair and maintenance manual or service manual etc.

3. Preparation

The sample tractor shall be well run in. The fuel tank, coolant and lubricant shall be filled to specified levels. Tyre inflation shall be in accordance to the pressure prescribed by the applicant in operation manual.

4. Specification sheet

The tractor manufacturer applicant shall furnish specifications of the tractor before start of the test. These shall be checked as far as possible by the testing station and reported in test report highlighting the deviations, if any.

5. Fuels and lubricants

Fuels and lubricants shall be selected from the range of products commercially available in the country where the equipment is tested. It shall conform to the minimum standards approved by the tractor manufacturer. If the fuel or lubricant conforms to a national or international standard, it shall be mentioned and the standard stated.

6. Auxiliary equipment

For all tests, accessories such as the hydraulic lift pump or air compressor may be disconnected, only if it is practicable for the operator to do so in normal practices of working otherwise they should remain connected and operated at minimum load.

7. Measuring instruments

Measuring instruments shall be inspected and calibrated before use.

8. Adjustments of the engine during the tests

The engine shall be thoroughly adjusted conforming to values given in printed manual/ specifications sheets, before the first test. These adjustments shall not be changed throughout the test.

9. Operating conditions

No corrections shall be made to the test results for atmospheric conditions or other factors. Atmospheric pressure shall not be less than 96.6 kpa. If this is not possible because of conditions of altitude, a modified fuel injection pump setting may have to be used, details of which will be included in the report. The pressure will be stated for each reading in the report. Stable operating conditions must have been attained at each load setting before beginning test measurements.

10. Ballasting

Ballast weights may be fitted. For tractors having pneumatic tyres, liquid ballast in the tyres may also be used: the overall static weight on each tyre (including liquid ballast in the tyres and a 75 kg weight representing the driver), and the inflation pressure shall be within the limits specified by the tyre manufacturer, except as specified for the five hour drawbar test. Inflation pressure shall be measured with the tyre valve in the lowest position

11. Repairs during tests

All repairs made during the tests shall be reported together with comments on any practical defects or shortcomings about which there is no doubt

12. Fuel consumption

a) When consumption is measured by mass

To obtain hourly consumption by volume and the work performed per unit volume of fuel, a conversion of unit of mass to unit of volume shall be made using the density value at 15 °C

b) When consumption is measured by volume

The mass of fuel per unit of work shall be calculated using the density corresponding to the fuel temperature at which the measurements was made. This figure shall then be used to obtain hourly consumption by volume and the work performed per unit volume of fuel using the density value at 15 °C for conversion from unit of mass to unit of volume

13. Testing two versions of the same tractor

If, at the request of the manufacturer, 2-and 4-wheel drive versions of one tractor are tested together the one version being modified to become the other, the same engine must be used in both and there must be no change in the transmission of power from the engine to the power take off.

14. Suspension of test

The test shall be suspended/stopped in any of the following case

- i) When any break down or abnormality occurs during the test and makes the normal execution of the subsequent test impossible.
- ii) When the applicant proposes to stop the test due to certain specific reasons

15. Retesting

Testing Station will retest a tractor model only if it has been modified so that its performance may be affected. If only the name has been changed, the station may certify that the test already reported applies to the new tractor as well.

16. Atmospheric conditions

a) Temperature

The ambient temperature for PTO performance test, belt pulley test and engine test shall be between 25-35 °C. The PTO test under high ambient condition shall be conducted at 45+2°C.

b) Atmospheric pressure

Minimum 96.6 Kpa is required during testing. If it not possible, modified fuel pump setting may be used and reported.

17. Permissible measurement tolerances

Rotational speed (rev./min.)	+0.5%
Time (sec.)	±0.2
Distance (m or mm)	±0.5%
Force (N)	+1.0%
Mass (Kg)	+0.5%
Atmospheric pressure (Kpa)	+0.2
Tyre pressure (Kpa or Kg/cm')	±5.0%
Hydraulic system pressure (Kpa)	±2.0%
Templeton of fuels etc.("C)	±2.0%
Wet and dry bulb thermometers ("C)	+0.5
Fuel consumption	
- Drawbar test (kg)	+2.0%
- PTO, belt and engine test (kg)	+1.0%*

3.3 Terminology

1- Length

The distance between the two vertical planes at right angles to the median plane of the tractor and touching its front and rear extremities. This is true when all parts of the tractors and in particular the components projecting at the front or rear are contained between these two planes. The removable hitch components at the front and rear are not included in the length.

2. Width

The distance between two vertical planes parallel to the median plane of the tractor, each plane touching the outermost point of the tractor on its respective side. This is true when all parts of the tractor, in particular all fixed components projecting laterally are contained between these two planes. The removable attachments like cage wheels etc. are not included in the width.

3. Height

The distance between the supporting surface and the horizontal plane touching the uppermost part of the tractor.

4. Track

The distance between the median planes of wheels on the same axle measured at the point of ground contact.

5. Wheel base

Horizontal distance between front and rear wheels measured at the centre of their point of ground contact.

6. Ground clearance

The distance between the firm horizontal supporting surface and the lowest point of the tractor.

7. Position of centre of gravity

The position of the centre of gravity is defined by

      a) Height above the supporting surface;
      b) Distance to right or left of the median plane of the tractor;
      c) Distance from the vertical plane passing through the line representing the track of the rear wheels, or sprockets.

8. Turning space diameter

The diameter of the smallest circle described by the outer most point of the tractor while executing its sharpest practical turn.

9. Turning diameter

The diameter of the circle described by the centre of tyre contact with the ground of the outermost wheel of the tractor while executing its sharpest practical turn.

10. Rated speed

The engine speed specified by the manufacturer for continuous operation at full load.

11. Engine power

The power measured at the flywheel or the crankshaft, when the engine is running at its rated speed.

12. Belt power

The power transmitted through a belt to a dynamometer for belt pulley work.

13. Power take-off power

The power measured at any shaft designed by the tractor manufacturer to be used for PTO work.

14. Drawbar power

The power available at the drawbar sustainable over a distance of atleast 20 metres.

15. Maximum drawbar pull

The mean maximum sustained pull which the tractor can maintain at the drawbar over a given distance, the pull being exerted horizontally and in the vertical plane containing the longitudinal axis of the tractor.

16. Specific fuel consumption

It is defined as the amount of fuel consumed per unit of power developed per hour. It is a clear indication of the efficiency with which the engine develops power from fuel. This parameter is widely used to compare the performance of different engines.

$$\text{Specific fuel consumption (SCF) (gm/hp-h)} = \frac{\text{Fuel consumed (gms/h)}}{\text{Horse power developed.}}$$

17. Specific energy

Work per unit volume of fuel consumed.

24

18. Slip

a) Belt slip is determined by the following formula:

$$\text{Belt slip (\%)} = \frac{100\,(n_0 - n_1)}{n_0} \qquad \ldots (3.1)$$

Where,

n_0 and n_1 are the number of revolutions per minute of the driven pulley at no load and under load respectively.

b) Slip of the driving wheels or track is determined by the following formula:

$$\text{Wheel or track slip (\%)} = \frac{100\,(n_1 - n_0)}{n_1} \qquad \ldots (3.2)$$

Where

n_1 = It is the sum of revolutions of all driving wheels or tracks for a given distance with load.

n_0 = It is the sum of the revolutions of all driving wheels or tracks for the same distance without drawbar load at the speed not exceeding 3.5 km/h.

Note: In the case of tractors having four driving wheels not mechanically locked together, the number of revolutions of each wheel should be separately recorded and the slip calculated for each wheel. If the results differ by more than 5%, the same should be noted and separately reported.

3.4 Basic measurements

The basic measurements which usually should be undertaken to evaluate the performance of an engine on almost all tests are the following:

A- Speed

A wide variety of speed measuring devices are available in the market. These range from a mechanical tachometer to digital and triggered electrical tachometers.

For accurate and continuous measurement of speed a magnetic pick-up placed near a toothed wheel coupled to the engine shaft is generally used. The magnetic pick-up will produce a pulse for every revolution and a pulse counter will accurately measure the speed.

B- Fuel consumption

The fuel consumed by an engine can be measured by determining the volume of flow of the fuel in a given time interval and multiplying it by the specific gravity of the fuel which should be measured occasionally to obtain accurate information.

Another method to determine fuel consumption is gravimetric method. In this method the time to consume a given weight of the fuel is measured. These measurements can be made automatic by the addition of suitable devices and are adequately accurate if proper care is taken in observing the readings.

Continuous flow meters like Flotron are also used to measure fuel consumption which give instantaneous readings. Such flow meters are very useful especially in testing of high horse power engine.

C- Smoke density

All the widely used smokemeters are basically carbon density (g/m^3) measuring devices. The meter readings are function of carbon mass in a given volume of exhaust gas.

A Bosch smoke meter is very popular. It has a calibrated chart for defining smoke density with bosch number. The basic principle in such smokemeter is that a fixed quantity of exhaust gas is passed through a fixed filter paper and the density of the smoke stains on the paper is evaluated optically. In a recent modification of this type of smokemeter, a pneumatically operated sampling pump and a photo-electric unit are used for the measurement of the intensity of smoke stain on filter paper.

Von brand smokemeter can give continuous readings of smoke density. A filter tape is continuously running at a uniform rate to which the exhaust from the engine is fed. The smoke stains developed on the filter paper are sensed by a recording head. The signal obtained from the reading head is calibrated to give smoke density.

D- Power measurement

Power is the most important measurement in the test schedule of engine. It involves the determination of the torque and the angular speed of the output shaft.

Observed power

The power obtained at the dynamometer without any correction for atmospheric temperature, pressure, or vapour pressure.

Corrected power

The power obtained by correcting observed power to standard conditions of sea level pressure (1.013×10^5 pa), temperature (15.5 "C) and zero vapour pressure.

3.5 Test items

1- Specifications checking

2- P.T.O. performance test

 A- Under normal (25-35 "C) ambient temperature condition.

  i) Test at varying speed
  ii) Two-hour test at maximum power
  iii) Test at standard pto speed
  iv) Test at varying load
  v) Test for pto which is not designed to transmit full power of **the engine.**

 B -Two-hours test at maximum power under high ambient temperature
  (45 ± 2"c) condition.

 C- Engine test for tractors (optional)

3- Belt-pulley performance test

4- Drawbar performance test

 A- Test for ballasted tractor

  i) Maximum power test
  ii) Ten-hours test

 B- Test for unballasted tractor

5- Hydraulic performance test

 A- Hydraulic lift test
 B- Hydraulic power test
 C- Maintenance of lift load test

6- Turning ability test

7- Position of centre of gravity and overturning angle

    A- Centre of gravity
    B- Overturning angle

8- Visibility test

9- Brake test

    A- Cold brake test
    B- Hot brake test
    C- Hand brake test

10- Air cleaner oil pull-over test

11- Noise measurement

    A- Noise at Bystander's position.
    B- Noise at operator's ear level

12- Mechanical vibration measurement

13- Field test

14- Water proof test

15- Assessment of power drop and wear

    A- Power drop test
    B- Wear test

3.6 Testing procedure

1- Specifications checking

The tractor shall be parked on a firm horizontal surface with fuel tank, coolant and lubricants filled upto the specified level. The tyre inflation pressure shall be adjusted to the value specified by the manufacturer for road work. The track setting shall be the one commonly used.

Specifications given by the manufacturer shall be verified and compared with the standard codes by the testing authority and reported highlighting deviation, if any.

2- PTO Performance test

The test shall be carried out on the tractor's main power take off shaft with the help of dynamometer as per details shown in Fig.3.1. The installation of fuel consumption meter shown in Fig.3.2 and for computation of specific gravity in Fig.3.3 for ultimately calculation of specific fuel consumption. The results of PTO tests can be summarized as shown in Appendix-III.

The torque and power values in the test report shall be obtained from the dynamometer without correction for losses in power transmission between the power take-off and the dynamometer. The shaft connecting the power take-off to the dynamometer shall not have any appreciable angularity at the universal joints and in any case not more than 2 degree.

Formulas for calculation:

a) $PS = \dfrac{2\pi NT}{75 \times 60} = \dfrac{TN}{716.2}$

PS = PTO horse power
T = Torque (Kgf.m)
N = Revolution (rev/min)

b) $F = \dfrac{S}{T} \times 3.6$

F : Fuel consumption (l/h)
S : Capacity (cc)
T : Time (sec)

c) $S = \dfrac{F}{PS} \times G$

S : Specific fuel consumption (g/hp-h)
F : Fuel consumption (cc/h)
G : Specific gravity (g/cc)
PS : Horse power of tractor

d) Fuel injection volume check (cc/stroke)

$$= \dfrac{\text{Fuel consumption (cc/sec)}}{\text{Engine speed (rpm)}} \times \dfrac{60}{\text{No.of cylinder}} \times 2$$

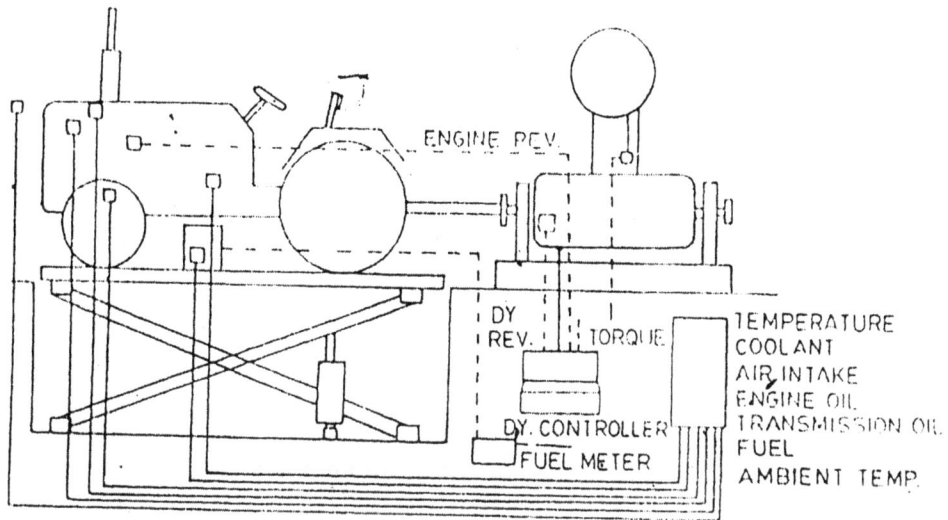

FIG. 3.1 INSTALLATION OF TRACTOR FOR PTO PERFORMANCE TEST

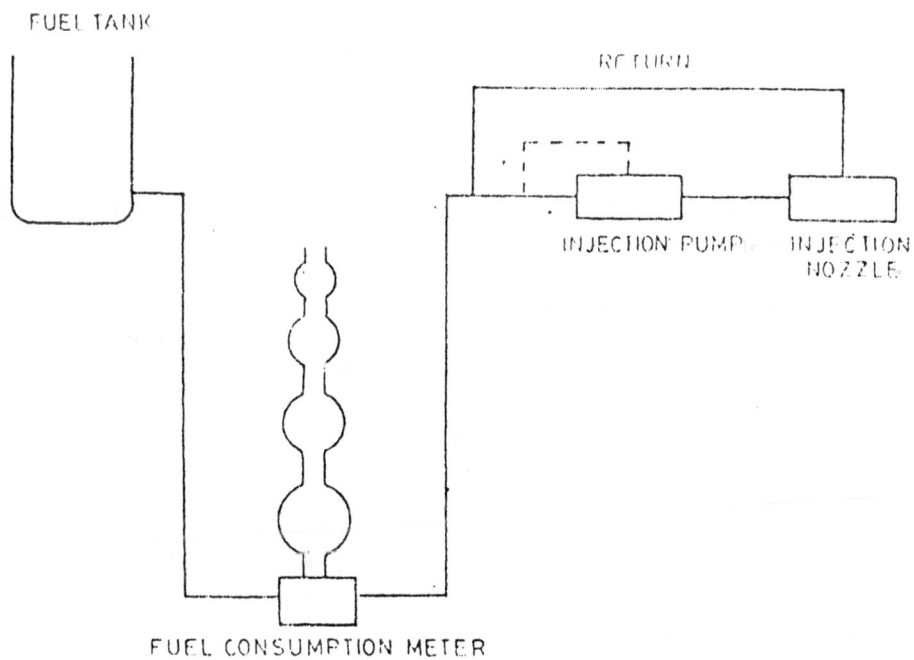

FIG. 3.2 INSTALLATION OF FUEL CONSUMPTION METER

30

Example:

Fuel consumption(cc/sec) : 50 cc/25 sec
Engine speed : 2000 rpm
No. of cylinder : 4

$$\text{Fuel injection} = \frac{50}{25} \times \frac{60}{2000 \times 4} \times 2 = 0.03 \text{ cc/stroke}$$

e) Governor control lever setting:

$$\frac{\text{Engine speed at no-load} \times \text{engine speed at standard PTO speed}}{\text{Engine speed at maximum power}}$$

Example:

Engine speed at no load: 2380 rpm.
Engine speed at the maximum power: 2200 rpm
Engine speed at standard PTO speed: 1900 rpm

$$\text{Governor setting} = \frac{2380}{2200} \times 1900 = 2055 \text{ rpm}$$

Performance curves

The test report shall include presentation of the following curves made for the full range of engine speeds tested:

a)   Power as function of speed (Fig.3.4)
b)   Equivalent crankshaft torque as a function of speed (Fig.3.5).
c)   Fuel consumption as a function of speed (Fig.3.6)
d)   Specific fuel consumption as a function of speed (Fig.3.7).
e)   Specific fuel consumption as a function of power (Fig.3.8)

Performance tests

A-   Under normal (25-35 ⁰C) ambient temperature condition

i)   Varying speed test

After the engine has reached the stabilized working condition, the fuel consumption,

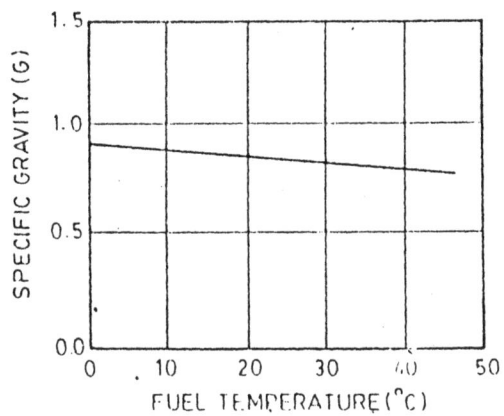

FIG.3.3 COMPUTATION OF SPECIFIC GRAVITY
(G) VALUES OF HIGH SPEED DIESEL
AT DIFFERENT FUEL TEMPERATURES(^{o}C)

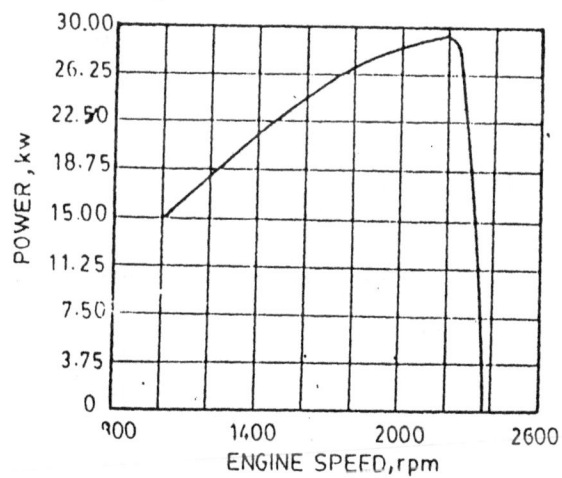

FIG.3.4 TYPICAL CURVE OF POWER Vs
ENGINE.SPEED

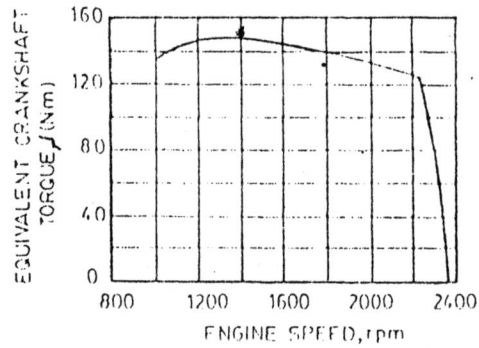

FIG.3.5 TYPICAL CURVE OF EQUIVALENT
CRANKSHAFT TORQUE Vs ENGINE
SPEED

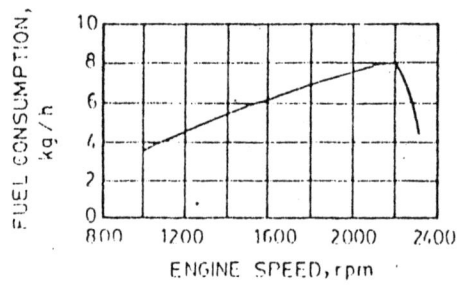

FIG.3.6 TYPICAL CURVE SHOWING HOURLY
FUEL CONSUMPTION Vs ENGINE
SPEED

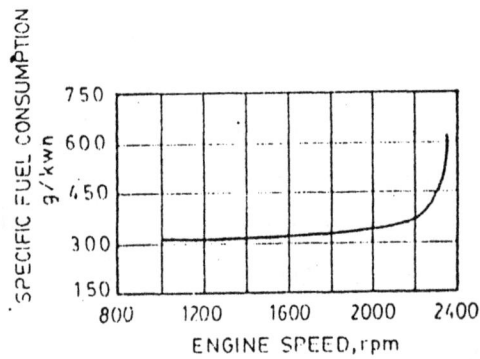

FIG.3.7 TYPICAL CURVE SHOWING SPECIFIC
FUEL CONSUMPTION Vs ENGINE SPEED

torque and power shall be measured as a function of speed by gradual loading up to full load. The minimum speed at which the measurements are made shall be at least 15% below the speed at maximum torque.

ii) Two hours test at maximum power

The tractor shall be operated at maximum power for two hours after the required warming up period for power to become stabilized. The governor control lever shall be placed in the position recommended by the tractor manufacturer for obtaining maximum power. A minimum of six readings at equal intervals of time shall be taken during the two hour test period. Recording of temperatures, pressures and other observations shall be made simultaneously.

The maximum power quoted in the report shall be the average of the readings taken during the two hours period. If the power variation is more than $\pm 2\%$ from the average, the test shall be repeated. If the variation continues to exceed $\pm 2\%$, it shall be reported in the report. The fuel and engine oil consumption shall also be determined during the two hours test.

iii) Performance at standard PTO-speed

If the engine speed recommended by the manufacturer for the test at maximum power does not give the standard pto shaft speed appropriate to the design of the shaft, the hourly fuel consumption, power and torque shall be measured during the varying speed test at the engine speed giving the standard pto speed.

iv) Varying load test

The power, torque and fuel consumption shall be measured for the load values in the sequence given below:

     a) 85 percent of load corresponding to the load at maximum power;
     b) Minimum load;
     c) 50 percent of the load given at (a);
     d) Load corresponding to the load at maximum power;
     e) 25 percent of the load given at (a); and
     f) 75 percent of the load given at (a).
     v) Special cases of tractors with a PTO shaft not designed to transmit the full power from the engine:

i) If the main PTO is unable pto transmit the full power of the engine, a two-hours test at a power specified by the manufacturer shall be conducted. If possible, a 20 percent overload shall be applied every 5 minutes for a period not exceeding 1 minute.

34

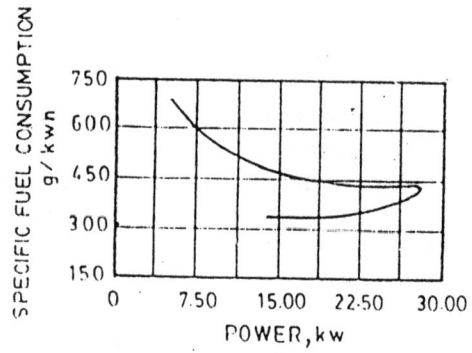

FIG.3.8 TYPICAL CURVE SHOWING SPECIFIC
FUEL CONSUMPTION Vs POWER

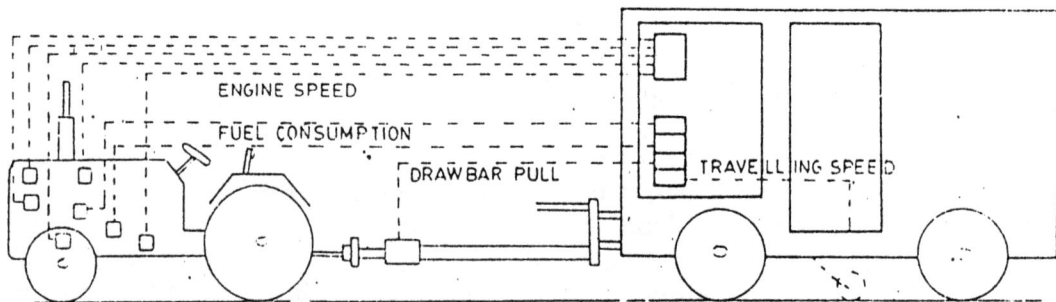

FIG.3.9 SCHEMATIC DIAGRAM FOR DRAWBAR PERFORMANCE TEST

ii) If the engine does not develop the 20 percent overload, the intermittent load shall be applied at the full engine power.

iii) In addition to above engine test shall also be conducted.

B- Test at high ambient temperature

Two hours test at maximum power shall be conducted under high ambient temperature condition. During the test, the temperature of the test cabin shall be maintained at $45+2^{\circ}$ C. The temperature of the test cabin can be raised artificially by using a suitable temperature raising device. During the test, the atmospheric pressure and relative humidity shall also be recorded.

C- Engine test for tractor

a) The test for the engine shall be carried out on a tractor model having no pto shaft or having a pto which is not mechanically connected to the engine, or is unable to transmit the full engine power.

b) The engine test may also be carried out on any tractor in addition to the PTO test, if desired by the manufacturer.

3- Belt pulley performance test

The tractor pulley shall be connected to the dynamometer through a flexible endless belt having appropriate power and torque transmission characteristics. The test is conducted under 25 to 35°C ambient temperature condition. The belt slip shall not exceed 2 percent and the tension necessary to prevent this shall be as small as possible. However, belt slip percentage shall be reported. Two hours maximum power test shall be conducted and results reported.

If the engine speed recommended by the manufacturer for maximum power test does not correspond to the standard belt speed of 15.75 ± 0.25 m/s, the performance test at the engine speed corresponding to the standard belt speed shall be conducted and evaluated.

4- Drawbar performance test

The tractor shall be coupled with the loading car and prepared for drawbar performance tests as shown in Fig.3.9. Tests shall be conducted with and without ballast conditions and results their of shall be summarized as shown in Appendix III.

Test conditions

The test at the drawbar shall be conducted according to the following regulations in order to provide reasonably comparable results.

  i) For tractors with pneumatic tyres, the test shall be carried out on a clean, horizontal and dry concrete or tarmacadam surface containing minimum number of joints;

Test of steel wheeled tractor and track-laying tractor shall be carried out on flat, dry and horizontal, mown or grazed grassland, or on a horizontal track having equally good adhesion characteristics.

A moving track (treadmill) may also be used subject to the condition that results produced are comparable to those obtained on the surfaces mentioned above.

  ii) During the test at the drawbar, the governor control shall be set for maximum power. Test shall not be made in gears in which the forward speed exceeds the safety limits of testing equipment.

  iii) The line of pull shall be horizontal. The optimum height of the drawbar under different ballast conditions shall be selected by the testing authority in consultation with the manufacturer. It shall be decided subject to the following limitations:-

  a) The height of the drawbar shall be such that when the tractor develops maximum sustained pull, the load exerted on the front wheels is sufficient to control the direction of travel.

  b) In case of wheeled tractors, the following formula shall apply:

$$H = \frac{0.8 \times W \times Z}{P}$$

Where:

  H = static height of the line of pull above the ground (mm).
  W = static load exerted by the front wheels on the ground (kg)
  Z = wheel base (mm)
  P = maximum drawbar pull (N)

  iv) At the beginning of the drawbar tests, the height of the tyre tread bars shall not be less than 65% of the height of the bars of the new tyres. The height of tyre tread bars shall be measured by a three-point gauge at the centre line of the tyre as far as possible.

  v) The measurements of drawbar pull, speed and slip shall be started only after the operational conditions are stabilized.

vi) The drawbar power shall be reported with reference to the gear used, mass of the tractor and its distribution on the front and rear axles, type of track, height of drawbar, tyre size and inflation pressure of front and rear tyres.

vii) For each gear at the forward speed and pull giving maximum power, the following item shall also be recorded. Drawbar pull, speed, wheel or track slip, fuel consumption, temperature of fuel, coolant and engine oil and atmospheric conditions.

Formulae for calculations

a) $$\text{Drawbar power (PS)} = \frac{\text{Drawbar pull (Kgf) x Traveling speed (m/s)}}{75}$$

b) $$\text{Wheel Slip (\%)} = \frac{100\,(N_o - N)}{N_o}$$

Where:

N_o = Traveling speed at no load.

N = Traveling speed at drawbar load

$$\text{where, } N_o = \frac{\text{Engine speed (rpm) x Distance traveled in one rev. of rear wheel (m)}}{\text{Gear reduction ratio x 60}}$$

Example: If engine speed = 2306 rpm

Distance traveled in one rev. of rear wheel = 4.53 m and

Gear reduction ratio = 56.78

Then :

$$N_o = \frac{2306 \times 4.53}{56.78 \times 60} = 3.07 \text{ m/s}$$

If N = 2.85 m/s

$$\text{Then slip} = \frac{100\,(3.07 - 2.85)}{3.07} = 7.2\%$$

$$\text{Gear reduction ratio} = \frac{\text{Engine speed (rpm)}}{\text{Rear drive wheel revolution (rpm)}}$$

Alternative method

$$\text{Wheel slip (\%)} = \frac{N1 - N2}{N1} \times 100$$

38

Where :

N1 = On load wheel revolution for 20 m length
N2 = No load wheel revolution for 20 m length

Performance curves

The test report shall include presentation of the following curves for each gear tested.

a) Drawbar power and forward speed as a function of drawbar pull. The curve should also indicate maximum sustained pull (see Fig.3.10).
b) Wheel or track slip as a function of drawbar pull (see Fig.3.11).
c) Specific fuel consumption as a function of drawbar pull in different forward gears (see Fig.3.12).

Performance test

A) Test for ballasted tractor

i) Maximum power test

The tractor shall be ballasted in accordance to recommendations of the tractor manufacturer within the load limit specified by the tyre manufacturer. The drawbar power at the rated engine speed of the ballasted tractor in each gear from the highest gear in which maximum power is limited by wheel slip of 15% or track slip of 7% to the gear immediately above that in which the highest maximum power is developed, shall be measured and reported.

ii) Ten hours test

The test of five hours each shall be conducted successively as follows:

a) Five hours test at 75% of the pull at maximum power:

The ballasted wheeled tractor shall be operated for 5 hours in a gear normally used for agricultural work, such as ploughing. The drawbar pull shall be 75% of the pull at maximum power in that gear. Data like power, speed, fuel consumption, temperature of fuel,coolant and lubricating oil and atmospheric conditions, shall be recorded and reported.

b) Five hours test at the drawbar pull coinciding with 15% wheel slip.

The ballasted wheeled tractor shall be operated for 5 hours at the drawbar pull giving 15% wheel slip measured during the test. The gear used shall be the fastest gear in which the required pull can be obtained when the engine is

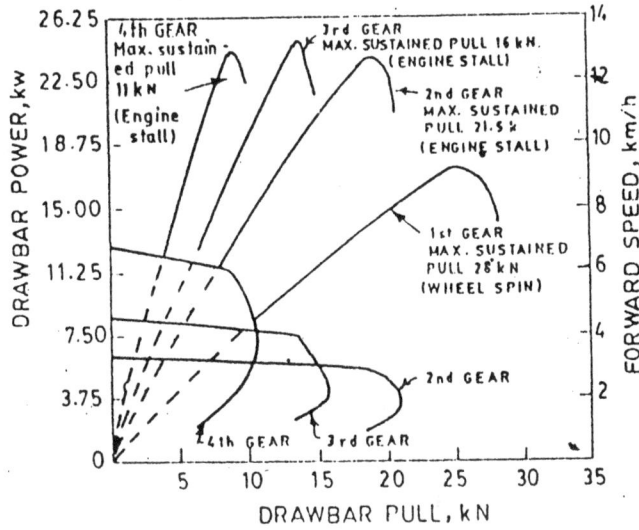

FIG. 3.10 TYPICAL CURVE SHOWING FORWARD SPEED
AND DRAWBAR POWER Vs DRAWBAR PULL

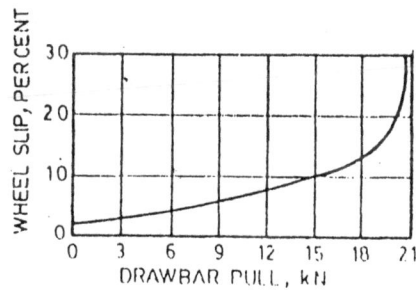

FIG. 3.11 TYPICAL CURVE SHOWING WHEEL
SLIP Vs DRAWBAR PULL

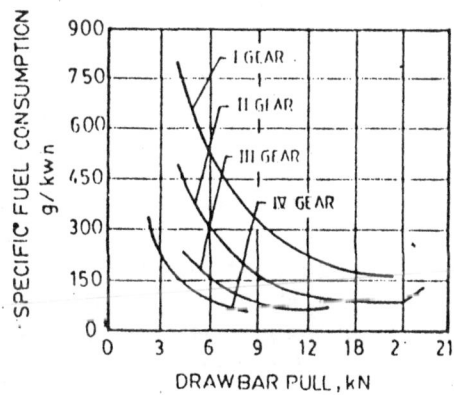

FIG. 3.12 TYPICAL CURVE SHOWING SPECIFIC
FUEL CONSUMPTION Vs DRAWBAR PULL

operating under the control of the governor. If necessary, supplementary ballast may be added to reduce the wear of the tyres and to facilitate control of the tractor. The drawbar pull, speed and atmospheric conditions shall be recorded and reported.

At the end of the ten hours test, the engine oil consumption shall be measured and reported in gm per hour.

B) Tests for unballasted tractor

Maximum power test only, shall be conducted on the same pattern as in the case of ballasted tractor.

5 Hydraulic performance test

The governor control lever shall be set for maximum power at rated engine speed. At the start of each test, the temperature of the hydraulic fluid in the tank shall be 65±5 °C. If this can not be achieved, owning to the presence of an oil cooler, the temperature measured during the test shall be stated in the test report. A pressure gauge shall be fitted immediately next to the external tapping of the tractor. To ensure the lifting capacity of the hydraulic lift to be adequate for effective practical use and also to allow for variation in the performance, the measured maximum performance should be reported, obtained at 90% of the pressure relief valve setting of hydraulic fluid pressure. The schematic diagram of hydraulic pump inline tester is shown in Fig.3.13.

Formula for calculations

$$\text{Hydraulic power (P.S)} = \frac{\text{Pump delivery rate (l/min.) x pressure(kgf/cm}^2)}{450}$$

Performance tests

A) Hydraulic lift test

The unballasted tractor shall be secured in the horizontal position in such a way that the tyres are not deflected by the reactive force of the power lift.

The lifting force shall be measured at the hitch points and on a frame attached to the three-point linkage. The frame shall have a mast height appropriate to the linkage category of the tractor as shown in Fig.3.14. Where more than one linkage category is specified, the one selected for test shall be at the option of the manufacturer in consultation with the testing authority.

Test frame:

Category	I	II	III
Mast height(mm)	460	510	560
L(mm)	683±1.5	825±1.5	965±1.5

The external force shall be applied to the frame at its centre of gravity and mass of the frame shall be added to this force. The centre of gravity shall be at a point 610 mm to the rear of the hitch points, and in line at right angle to the mast and passing through the middle of the line joining the lower hitch points.

The side linkages shall be adjusted to satisfy the vertical movement and the minimum height at the hitch point as shown in Fig-3.15

Category	I	II	III
Minimum height (H) (mm)	200 ≥	200 ≥	230 ≥
L(mm)	560 ≤	600 ≤	685 ≤

The lifting force available shall be measured and reported at a minimum of six points approximately equally spaced throughout the range of movement of the lift. At each point, the force shall be the maximum which can be exerted against a static load. The following measurements should be made and reported:

1. The maximum force at the hitch point and the frame measured and corrected to those values corresponding to a hydraulicpressure equivalent to 90 percent of the pressure relief value setting.

2. The pressure corresponding to 90 percent of minimum relief valve pressure setting as specified by the manufacturer.

3. The vertical movement of the point of application of the force.

4 The height of the lower hitch point above the ground in its lower most position and without load;

5 The lifting angle of the mast over the full range of the lift.

6 The temperature of the hydraulic fluid in the tank during the test.

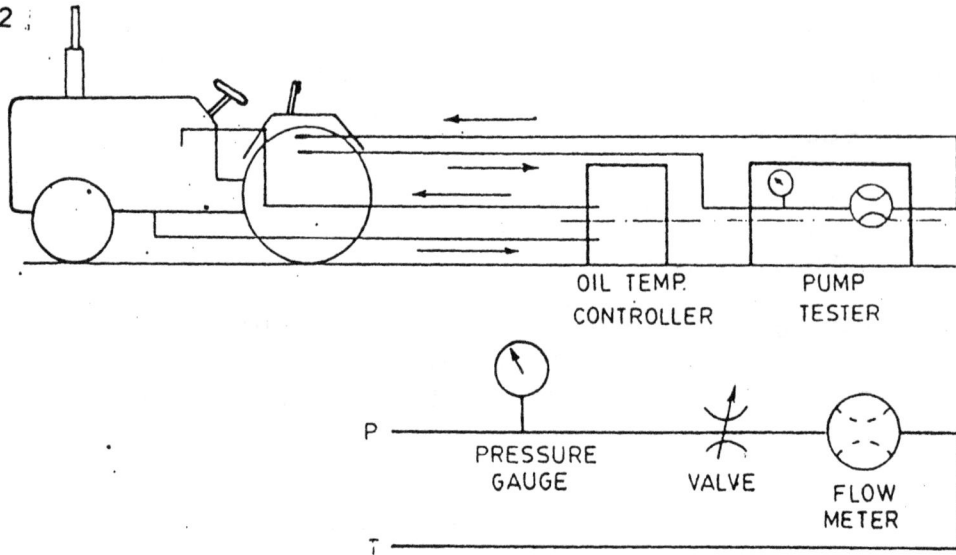

FIG.3.13 SCHEMATIC DIAGRAM OF HYDRAULIC PUMP IN LINE TESTER

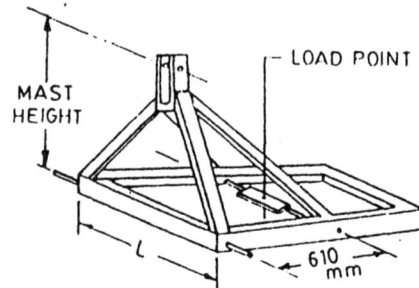

FIG.3.14 TYPICAL DIMENSIONS OF MOUNTED FRAME

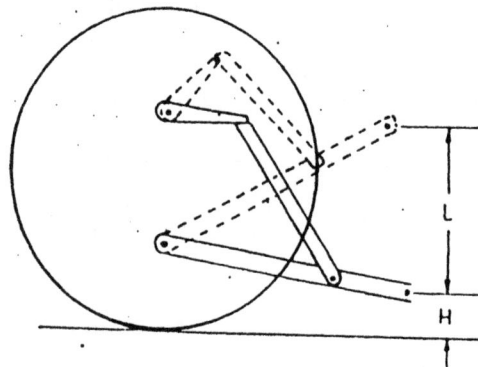

FIG.3.15 OPERATIONAL REQUIREMENT OF SIDE LINKAGE

7) The calculated moment around the rear axle,resulting from the maximum external lift force at the frame, exerted through the full range of movement.

8) A mass equal to the force as obtained during test shall be suspended from a point 610 mm to the rear of the hitch point. The time of the lift through the full range of the lift at rated engine speed shall also be reported.

B) Hydraulic power test

The following measurements shall be made and reported:

i) The opening pressure of the relief valve, and, if not possible to measure, then manufacturer's setting.

ii) The pump flow rate at rated engine speed and at minimum pressure.

iii) The pressure sustained by the open relief valve with the pump stalled and in the case of a closed-centre system with pressure-compensated variable delivery pump.

iv) The opening and closing pressure of the unloading valve in the case of a closed-centre system having an accumulator.

v) The pump delivery rate at every 20 kgf/cm² from low to high pressure.

vi) The hydraulic power available at any hydraulic tapping and the corresponding flow rate, at 90% of relief valve pressure setting, or for continuous operation of an external hydraulic motor, whatever pressure is specified by the manufacturer.

vii) The hydraulic pressure at maximum power.

viii) The temperature of the hydraulic fluid in the tank during test.

C) Maintenance of lift load test

A vertical force to the frame equal to 90% of the maximum force exerted throughout the full range of movement shall be applied at its centre of gravity. When the hydraulic lift is at the uppermost position and the control lever in the 'raise' position, the engine shall be stopped and following measurements shall be made and reported.

a) The force applied to the frame.

b) The decrease in the vertical height at 5 minutes interval over a period of 30 minutes.

6. Turning ability test

The tractor shall be tested either on a test track as in the drawbar test or on a horizontal and compact area. The track setting shall be one commonly used for general purpose and be stated. The tractor shall be unballasted and moving slowly at a speed of 1.5 to 2 km/h. The test shall be made turning the tractor to right and left with and without using the steering brakes. After each turn (right or left) the following shall be measured and recorded as shown in Fig. 3.16.

  a) Minimum turning diameter.

  b) Minimum clearance diameter.

7. Position of centre of gravity and over turning angle

These shall be determined with fuel tank, oil sump full upto the recommended levels and driver replaced by a weight of 75 kg on the driver's seat, with and without ballast.

A- Centre of gravity

The point of centre of gravity can be located if we know height from ground, horizontal distance from centre of rear axle and horizontal distance from the median plane of the tractor. Arrangements for measurement of these dimensions are shown in Fig.3.17.

Formula for calculation

$$l_1 = \frac{W_1 \times l}{W}$$

$$h = r_1 + \frac{l_1 - (l'_1 / \cos\theta)}{\text{Tan }\theta}$$

$$q = \text{Tan}^{-1} \frac{(r_1 - r_2)}{l} + \frac{\text{Tan}^{-1}(\gamma'_2 + y - \gamma'_1)}{l'}$$

$$l_2 = \frac{T(W/2 - w)}{W}, \qquad T = \frac{Tr + Tf}{2}$$

l_1 = Horizontal distance from centre of rear wheel axle to centre of gravity

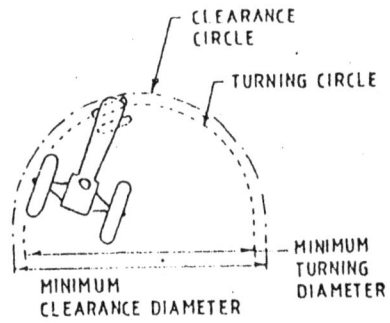

FIG.3.16 TYPICAL ARRANGEMENT FOR TURNING
ABILITY

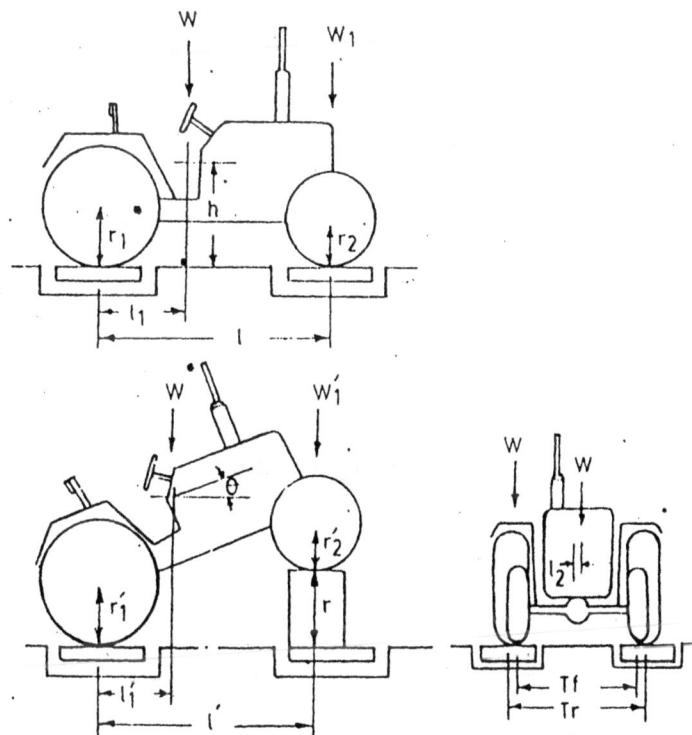

FIG.3.17 ARRANGEMENT FOR POSITION OF CENTRE
OF GRAVITY

h = Height of centre of gravity above ground

l_2 = Horizontal distance from the median plane of the tractor to the centre of gravity

W_1 = front axle weight at horizontal position

l = wheel base

r_1 = height of centre of rear axle above ground at horizontal position

l' = horizontal distance from centre of rear wheel axle to center of front wheel axle with front wheel lifted up

θ = angle of inclination.

l' = horizontal distance from center of rear wheel axle to center of gravity with front wheel lifted up.

r_2 = height of center of front axle above ground at horizontal position

$γ'_1$ = height of center of rear axle above ground with front wheel lifted up.

$γ'_2$ = height of center of front axle above ground with front wheel lifted up

y = lifting height of front wheel above ground

T = Track

Tf = Front track

Tr = Rear track

w = right side or left side weight of the tractor

W = Wt of tractor

B   Overturning angle:

$$θ = \tan^{-1} \frac{(T \pm 2l_2)}{2h} \quad \text{where } T = \frac{Tr + Tf}{2}$$

l_2 = horizontal distance from the median plane of the tractor to center of gravity

T = Track

Tf = Front track

Tr = Rear track

h = height of center of gravity above ground

Visibility test

The unballasted tractor shall be parked on a level and horizontal surface. An artificial light i.e. an electric bulb will be placed at a height of 720 mm from the centre of driver seat referred as an operator's vision and 75 kg weight will be placed on driver's seat, referred as weight of operator. The line separating the visible area from the non-visible shall be marked, measured and graph drawn as shown in Fig.3.18.

9. Brake test

The test shall be carried out on a plane, horizontal, dry and clean pucca surface. The settings and condition of the brake components including tyre pressure on the road shall be as per manufacturer's recommendation and specifications. The test shall be carried out at highest forward speed or at 25 km/h, whichever is lower. The following tests shall be conducted:

A- Cold brake test

The brakes shall be considered as 'cold' when its temperature is lower than 100"C. If it is not practicable to measure the temperature of the brakes, test shall not be started within one hour of the last severe braking and a pause of 5 minutes shall be allowed between each braking test. The brakes shall be applied with successively increasing force on the pedal. until the force giving the shortest stopping distance is found. At this point,the following shall be recorded:

      a) Deceleration
      b) Stopping distance
      c) Force exerted on the brake pedal
      d) Braking efficiency

In addition, the force exerted on the brake pedal to achieve a maximum braking deceleration of 2.5 m/s^2 shall be recorded.

B- Hot brake test

To warm up the brakes, the tractor shall be towed with the clutch disengaged but partially brakes engaged to develop pull equal to one tenth of the ballasted mass

48

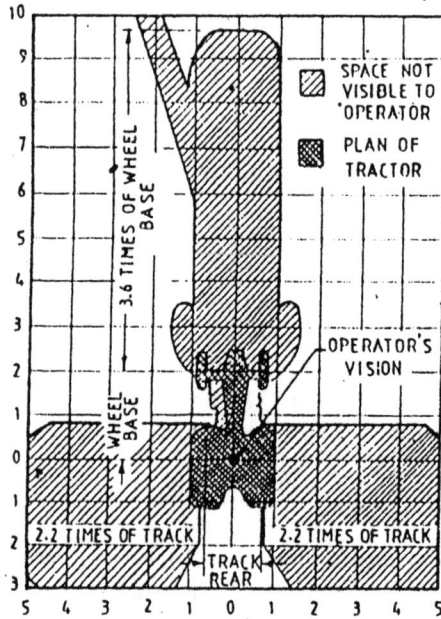

FIG.3.18 TYPICAL GRAPH SHOWING VISIBILITY FROM
DRIVER'S SEAT.

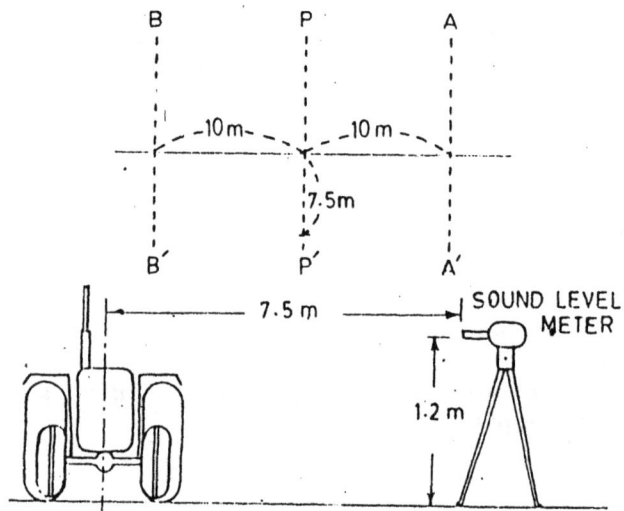

FIG.3.19 NOISE MEASUREMENT EMITTED
BY TRACTOR

of the tractor under test, over a distance of one kilometre, by an other tractor equipped with an engine speed tachometer. at 80 percent of the speed of the tractor selected. The towing tractor should be able to maintain constant speed and be of sufficient mass so that it does not slip. Before starting this warming up, the brakes should be cold. After warming up the data similar to cold brake test shall be recorded.

C) Hand brake test

The tractor shall be placed out of gear on a 16% slope facing first up and then down the slope. The parking brake shall be applied and the force applied on hand control lever shall be recorded. This force should not exceed 400 N. At this force, if wheel does not move then the hand brake is considered effective otherwise not.

10. Air-cleaner oil pull-over test

The tractor shall be placed on a level laid ground. The air-cleaner shall be cleaned and filled with oil of viscosity recommended by the manufacturer, upto the marked level. The engine shall then be operated at full governor speed for 15 minutes. This shall be followed by sudden accelerations and decelerations made after every 30 seconds for a period of 15 minutes. The air-cleaner assembly shall be weighed before and after the test. The loss of mass of oil, shall be calculated and reported.

If there is no oil pull over with the tractor in the level position, the following additional tests shall be carried out.

In case of wheeled tractor, the tests shall be repeated with tilt angle of 15" to either side and then 15 " forward and backward in relation to the direction of travel of the tractor. In case of crawler tractor, the forward and backward tilt angle shall be 30".

11. Noise measurement

A sound level meter is used for noise measurement in decibel dB(A). The test site shall be an open space of radius not less than 50 m and free from any object like building, rocks etc. The ambient temp. shall not exceed 35"C. The test shall be conducted preferably during night so that external noise does not affect the results. The following tests are conducted

A- Noise at bystander's position

The microphone shall be placed at 1.2m above ground and a distance of 7.5 m from the axis of forward movement of the vehicle at perpendicular to PP' on each side for measurement. Two lines AA' and BB' parallel to the line PP' and situated respectively 10 m in front and behind this line, shall be marked on the test track (Fig.3.19).

Tractor shall be driven at stabilized speed corresponding to three quarters of max. engine speed using the top gear. When the front of the tractor reaches line AA', the throttle shall be quickly fully opened. The lever shall be held in this position until the rear of vehicle has passed line BB' and then brought to the minimum position as quickly as possible. Measurements shall be carried out on each side of the vehicle.

B. Noise at operator's ear level

The noise shall be measured for the unballasted tractor in a sufficiently silent and open zone. The microphone fitted with the helmet frame shall be placed at 5 cm away as shown in Fig.3.20. The governor control lever shall be set for maximum power at rated speed of the engine.

Sound level measurements are performed in all gears and at the gear which gives the nominal speed. Starting nearest to 7.5 km/h with no load, the load shall be applied corresponding to 15% slip in case of wheel type and 7% slip in case of track type tractor. However, if engine stall is the limiting factor then the load corresponding to max. drawbar power obtained shall be applied.

12. Mechanical vibration measurement

The amplitude of mechanical vibration of those assemblies and components of the tractor which are functionally important,from operator's comfort and durability point of view shall be measured with the help of suitable vibration-measuring device,

The tractor shall be parked on a level concrete surface and operated at rated speed at no-load and at load corresponding to 85 per cent of the load at maximum PTO power. The maximum horizontal displacement(HD) and vertical displacement(VD) in microns shall be measured with the help of vibration meter.

13. Field test

The field tests with commonly used implements namely disc plough, mould board plough,cultivator and disc harrow shall be conducted to assess practical performance of the test tractor. A test duration of 50 h with each of these implements is recommended. In addition, puddling test of 50 h may also be conducted if the tractor has been declared suitable for wet land operations. Various performance parameters like rate of work,fuel consumption, quality of work, etc. shall be evaluated based upon observations recorded during such tests. The details are given in Chapter-IV.

14. Water proof test

The test shall be carried out on the tractor suitable for wet land cultivation. The object is to ascertain the dust and water proofing, mainly of wheel and brakes. The

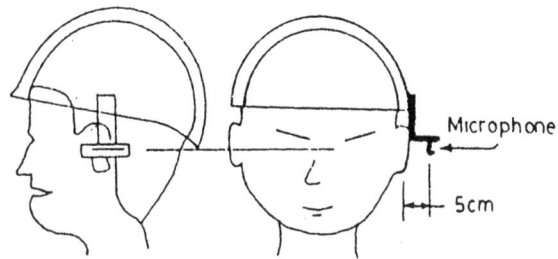

FIG. 3.20 ARRANGEMENT FOR NOISE MEASUREMENT
AT OPERATOR'S EAR LEVEL

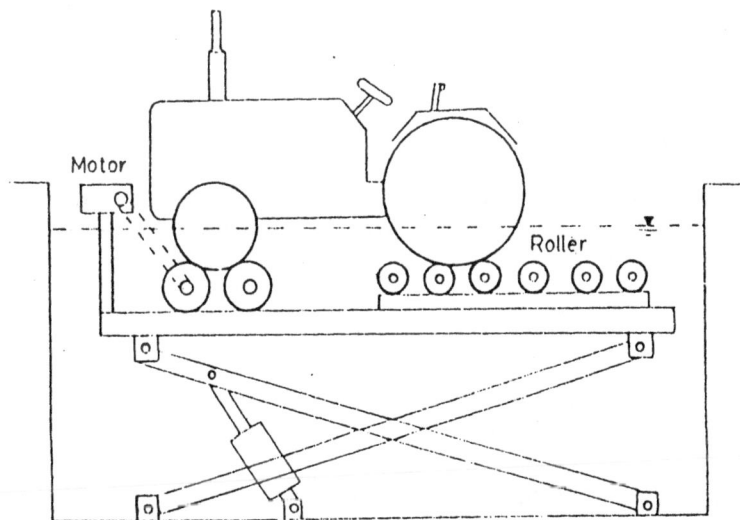

FIG. 3.21 SCHEMATIC DIAGRAM FOR WATER-PROOF TEST

tractor shall be operated at the speed approx. 6 km/h on the test bed as shown in Fig 3.21 for a period of 5 hours. The water level in the bed shall be adjusted to the centre of front wheel. After the test, axle and brakes shall be dismantled and inspected for any entry of mud or water.

15. Assessment of power drop and wear test

A- Power drop test

After conducting all laboratory and field tests, duration of which shall not be less then 300 hours, a two-hour maximum power test shall be carried out to see any drop in power. While conducting this test, no special adjustment or change of part shall be made except normal adjustment as specified by the manufacturer.

B- Wear test

i) Cylinder bore

The cylinder bore shall be measured on the thrust side and perpendicular to it at the top, middle and at the bottom position of the liner. The wear (%) shall be calculated and reported.

ii) Piston diameter

The piston diameter shall be measured on the thrust side and perpendicular to it at the top above the gudgeon pin and at the skirt. The wear (%) shall be calculated and reported.

iii) Ring end gap

The ring end gap for all compression and oil rings shall be measured at the top, middle and at the bottom position of the liner. The wear (%) shall be calculated and reported.

iv) Ring groove clearance

The ring groove clearance shall be measured with piston and oil rings. The wear (%) shall be calculated and reported.

v) Clearances of main and big end bearing

The radial and axial clearance of main and big end bearings shall be measured. The radial clearance shall be measured after tightening the crank shaft bolts with the torque specified by the manufacturer. The wear (%) shall be calculated and reported.

vi) Valves, guides and tappets

The valve shall be inspected for carbon deposition on the stem, overheating sign of pitting of the seats if any. The timing gear cover shall be opened and backlash between each pair of the meshing gears shall be measured, calculated and reported.

vii) Clutch

The clutch shall be opened and inspected for wear of the lining and pressure plate, condition of the clutch-release bearing, pilot bearings, springs and fingers. The clutch housing shall be inspected for the entry of dust, water, etc.

viii) Gear box

The top cover of the gear box shall be opened and inspected for visual wear and damage to the gear teeth. The backlash between each pair of the meshing gears shall be measured. The wear percent of backlash shall be calculated and reported. If the tractor has been used for puddling test, the entry of water and mud inside the gear box housing shall also be checked.

ix) Brakes

The brake housing shall be opened and inspected. The wear of brake lining shall be determined by measuring the thickness of lining and wear (%) of the lining shall be calculated and reported. If the tractor has been used for puddling test the entry of water and mud in the housing, shall also be inspected and reported.

x) Front axle

The king pin and stub axle shall be dismantled and inspected for wear of king pin and bushes. The condition of thrust bearings, bearings and seals for stub axle, and king pin shall also be examined for entry of dust, water, etc. For track type tractors, wear of sprocket, pin, grouser plate, idler, etc., shall be measured and reported.

xi) Dynamo and starter

These shall be dismantled and inspected for entry of water and dust. The condition of bearings shall also be examined and reported.

Rate of wear

The rate of wear shall be calculated on the basis of the initial average value when the engine is new and the maximum permissible wear indicated by the tractor manufacturer in the instruction manual.

Example:

Calculate the wear percentage, if the piston ring gap in the initial setting is prescribed as 0.3 to 0.5 mm. The manufacturer recommends that the piston rings should be changed when the gap reaches 1.5 mm. The measured value after the test was found to be 1.2 mm.

Average initial wear = 0.3 + 0.5 = 0.4 mm

$$\text{Observed wear} = \frac{(1.2 - 0.4)}{(1.5 - 0.4)} \times 100 = 72.7\%$$

3.7 Interpretation of tractor test results

PTO performance : measured in the laboratory with the pto of the tractor coupled to a dynamometer.

Maximum power : It is the maximum power, measured during two hours maximum power test at natural ambient temperature of 25-35 degree C, available at the tractor pto with the governor control fully open.

Power at standard pto speed: It is the maximum power available for pto work at the standard pto speed of 540/1000 rpm at natural ambient temperature of 25-35 degree C.

Torque back up ratio: It gives an indication of how much the engine torque increases to maximum (ratio of maximum torque and torque at maximum power) and hence the ability of the machine to keep working as the engine speed pulls down due to increased load requirements over and above the maximum power.

Specific fuel consumption (SFC) at maximum power : It is the mass of fuel consumed per unit work measured during two hours maximum power test (at natural ambient temperature of 25-35 degree C) indicates the fuel efficiency at that power level.

Drawbar performance : It is measured on a specially constructed concrete test track under both unballasted and ballasted conditions with the governor control fully open.

Maximum pull : The maximum pull available at the drawbar at 15% wheel slip of the tractor or Refers to pull corresponding to rated engine speed in case of tractors where 15% wheel slip level is not reached because of their high weight/power ratio.

Hydraulic performance : It is measured in the laboratory on a special designed test rig.

Maximum hydraulic power : It is the power available to drive the hydraulic ram of the tractor to lift load or to drive external hydraulic rams on implements and machines with the governor control fully open.

Flow rate at maximum power : It indicates the speed at which the hydraulic ram of the tractor or.implement and machines can be operated.

Lift capacity at standard frame : determines the force that can be applied by ram on an implement or a machine.

Brake performance : It is measured on the standard concrete test track.

Maximum stopping distance : The stopping distance obtained at an applied force of 600 N (61 kg. approximate) on the brake pedals and the tractor initially running at its maximum achievable speed.

Force for deceleration of 2.5 m/s^2 : to check the effort to produce the braking performance to be within 600 N.

Parking performance : the effectiveness of the parking braking device (at force not exceeding 600 N at brake pedal and 400 N at hand brake respectively) to hold the tractor stationary when facing up or down under unballasted conditions at 12% gradient with trailer having gross mass equal to the mass of tractor of 3 tonnes whichever is minimum and at 18% gradient under ballasted conditions.

Noise level: It is measured in the open on a concrete track, both at the Bystander's position and at the operator's ear level position and should not exceed 90 dB(A) as specified by ILO (International Labour Organization).

Mechanical vibration : It is the amplitude of average mechanical vibration of assemblies and components measured at the tractor. which are functionally important with the engine running at its rated speed both on load and unload conditions and should not exceed 100 microns.

Field performance : It is carried out in the field for a duration of atleast 50 hours with each implement. It should be noted that the performance and fuel consumption varied over the operating range of the engine should be considered and not a single figure be taken as representative for all conditions.

Chapter 4

TESTING AND EVALUATION OF TILLAGE MACHINERY

4.1. Introduction

Manufacture of the tillage implements in India is reserved for small scale sector. These units are mostly located in the states of M.P., U.P., Punjab, Haryana, Bihar and Karnataka. About 70% of these units have an investment of less than Rs.2 lakhs and only about 4% over Rs.5 lakhs, in plant and machinery. With such a small investment, there is no facility for design, development testing and quality control. These deficiencies result in more energy requirement, frequent breakdowns, less output and unsatisfactory performance. Though, the Bureau of Indian Standards has brought out about 300 standards in the area of farm machinery, only 650 product licenses have been granted by them as against a total number of about 17000 manufacturers in the country.

Besides, there is a growing interest in minimum tillage system with a view to reducing production cost, improving soil conditions and saving energy input.

The main objectives of quality tillage implements are:

1. To minimize the mechanical power and labour requirements.
2. To conserve the moisture and reduce soil erosion.
3. To perform the operations necessary to optimize soil tilth.
4. To maximize the field efficiency.

Therefore, testing and evaluation of tillage implements under the laboratory and actual field conditions is the only way of recognizing their quality and output. This will also help Govt. agencies in controlling the undeserving units, improving the deficiencies in existing implements, standardization of critical components and saving the energy through adoption of energy efficient implements. For example, a single ploughing with soil inversion plough gives better tilth than a four times ploughing with country plough, in addition to saving in cost of cultivation and saving in time for seed bed preparation for sowing of the succeeding crop.

4.2. Type of tests

A. Laboratory Tests

a. Checking of specifications
b. Hardness of critical components
c. Chemical analysis of soil engaging components
d. Wear of soil engaging components

B. Field Tests

a. Rate of work
   i) Width of cut
   ii) Effective field capacity
   iii) Field efficiency

b. Quality of work
   i) Depth of cut
   ii) Soil inversion
   iii) Soil pulverization

c. Draft measurement
d. Fuel consumption
e. Soundness of construction
f. Ease of adjustments and operation

4.3. General conditions

1. Selection of test samples

The test sample which is in commercial production shall be selected at random from the works of the manufacturer by a representative of testing institute. However, a prototype machine may be submitted directly to the testing institute. The method of selection should be given in the test report.

2. Acquaintance with the test sample

The testing team should get fully acquainted with the construction and operational features of the equipment/machine to assess its real function and performance. The technical literature/information/manufacturing drawings etc. should be thoroughly studied. All adjustments should be made as per the manufacturer's recommendations. It should also be ensured that test operator is fully conversant with the operation of machine.

3. Instrumentation

The reliability of testing data, to a great extent depends upon the accuracy of instrumentation. For proper evaluation of test sample, it is necessary that various testing institutions, test a machine on a common yard stick. Therefore, the measuring instruments should have accuracy as specified below:

Time(s)	± 0.2 sec
Distance	± 0.5%
Force (kgf)	± 2.0%

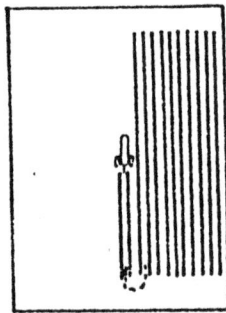

CONTINUOUS PATTERN
TURN STRIPS/EACH END
AT

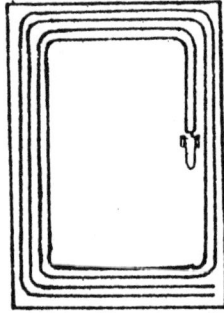

CIRCUITOUS PATTERN
ROUNDED CORNERS

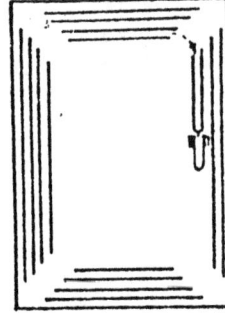

CIRCUITOUS PATTERN
TURN STRIPS AT CORNER
DIAGONALS

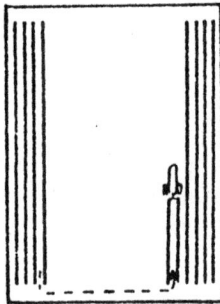

HEADLAND PATTERN,
FROM BOUNDARIES

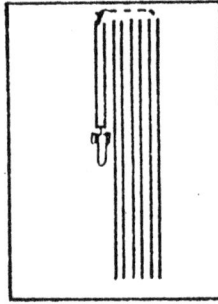

HEADLAND PATTERN,
FROM BACK FURROW

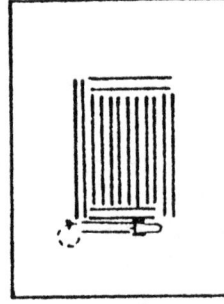

CIRCUITOUS PATTERN 270°
TURNS FROM BOUNDARIES
OR CENTRE

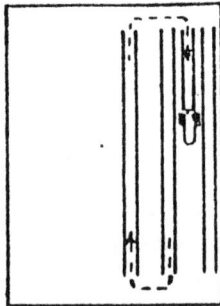

OVERLAPPING
ALTERATION PATTERN

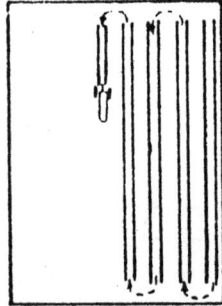

STRAIGHT ALTERATION
PATTERN

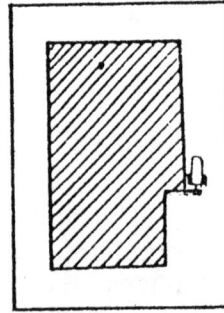

CIRCUITOUS PATTERN
SQUARE CORNERS

FIG.4.1 FIELD OPERATION. . PATTERNS

Mass (kg)   ± 0.5%
Rotational speed(rpm) ± 0.5%

4. Selection of test plot

Tillage equipment/machinery gives best performance in rectangular fields. The test plot should be rectangular with side lengths in the ratio of 2:1 as far as possible. If the field is irregular, then a rectangular test plot should be marked for conducting the test. The other portion of the field could be used for initial setting-up and adjustment of the equipment. As far as possible, the test plot selected should not have any previous tillage treatment after the last crop harvested.

5. Field operational pattern

Field capacity and field efficiency of an implement are affected by field operational pattern which is closely related to the size and shape of the field, the kind and size of implement. The down time (non-working time) should be eliminated with adoption of appropriate field operational pattern. Common field operational patterns for rectangular field are shown in Fig. 4.1.

6. Speed of operation

To calculate the speed of operation, two poles 20 m apart are placed approximately in the middle of the test run. On the opposite side also two poles are placed in a similar position and 20 m apart so that all four poles form corners of a rectangle. The speed will be calculated from the time required for the machine to travel the distance of 20 meter between the assumed line connecting two poles on opposite sides AC and BD as shown in Fig.4.2. The easily visible point of the machine should be selected for measuring the time.

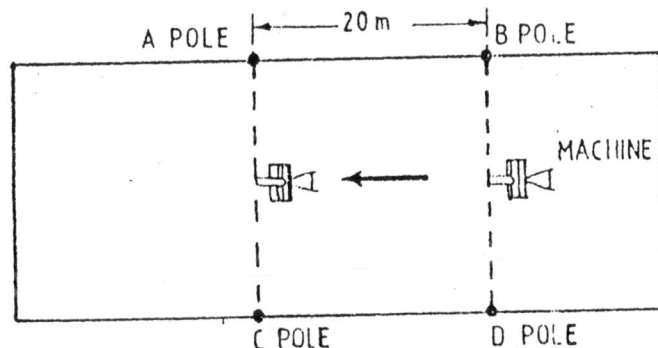

FIG.4.2 SPEED OF OPERATION MEASUREMENT

7. Wheel slip

Tractor drive wheel normally slip in all field operations. The distance covered by the tractor in a given number of drive wheel revolutions decreases with the wheel slip. Therefore, in case of tractor or power tiller operated implement the wheel slip will affect the speed of operation and thereby the effective field capacity of the implement. The wheel slip is determined as under and expressed in percentage.

$$\text{Wheel slip (\%)} = \frac{N_L - N_O}{N_L}$$

Where NL and NO are total no. of revolutions at load and at no load respectively for the marked test run.

8. Duration of test

The test sample should be operated under different soil and surface conditions for a minimum period of 50 h to establish its performance. Each test should be of minimum 3 hours. However, for exhaustive testing to establish soundness of construction and durability, minimum testing of 200 hours is recommended.

9. Field parameters

Various parameters to define soil characteristics and surface conditions of the test plot as specified below, should be observed and recorded:

>   a. Location of test plot
>   b. Size of test plot
>   c. Last crop grown
>   d. Detail of previous tillage operation, if any.
>   e. Topography of field
>   f. Type of soil
>   h. Bulk density of soil
>   i. Cone index of soil

Moisture content for soil is computed on dry weight basis. For measurement of soil moisture, take core samples of wet soil from atleast three different locations randomly selected in the test plot. Weigh the sample and record the weight. Place the sample in a hot air oven maintained at 105° for atleast 8 hours. At the end of 8 hours, cool the sample in a desiccator and weigh again. Calculate the soil moisture using the following formula :

$$\text{Soil moisture (\% dry wt. basis)} = \frac{\text{Wt. of wet soil sample - wt. of dry soil sample}}{\text{Wt. of dry soil sample}} \times 100$$

$$= \frac{\text{Weight of moisture in soil sample}}{\text{Weight of dry soil sample}} \times 100$$

Bulk density of soil is defined as the mass after oven drying of soil per unit volume. For measurement of bulk density of soil, take cylindrical core sample from atleast three different locations selected randomly in the test plot. Measure the diameter and length of cylindrical soil sample. Keep the core sample in the oven, maintained at 105" for atleast 8 hours. At the end of 8 hours, take out the sample from the oven and cool it in a desiccator and weigh again.

$$\text{Bulk density of soil (g/cc)} = \frac{M}{V} = \frac{4M}{\pi D^2 L}$$

Where M= Mass of oven dry core sample (g)
V= Volume of cylindrical core sample (cc)
D= Diameter of cylindrical core sample (cm)
L= Length of cylindrical core sample (cm)

Cone index also provides an indication of soil resistance and is expressed as force per square centimeter required for a cone of standard base area to penetrate into soil to different depths. Cone index for the same soil varies with the cone apex angle, area of cone base and depth of penetration. Apex angle and area or diameter of cone used should be given in the report. Profile of cone index to depth and/ or mean value up to working depth of implement should also be reported.

4.4. Testing procedure
A. Laboratory test

a- Checking of specifications: The specifications of the equipment/machine and of main components should be checked and verified as against values furnished by the applicant given in BIS document. The variation if any, should be highlighted in the report.

b- Hardness : The hardness of various critical components should be measured and compared with the relevant Indian Standards.

c- Chemical analysis: The chemical composition of critical soil engaging components should be determined and reported as shown in Appendix-IV.

d- Wear test : The mass of critical soil engaging components should be determined

62

before and after the field tests to assess wear rate.

B. Field test

Various following parameters are observed, evaluated and summarized as shown in Appendix-V (a) & V (b).

a) Rate of Work

i) Width of cut

For determining width of cut, average of 5 runs should be taken. The measurement of composite width should be taken at minimum 5 equidistant places in the direction of travel and average working width should be determined.

In case of disc harrow, width of cut may also be determined with the help of following formula:

For single action disc harrow

$$W = \frac{0.95 \, NS + 0.3D}{1000}$$

For double action disc harrow (offset type)

$$W = \frac{0.95 \, NS + 0.6D}{1000}$$

For tandom disc harrow

$$W = \frac{0.95 \, NS + 1.2D}{1000}$$

Where :

W = width of cut (m)
N = number of disc spacing
S = disc spacing (mm)
D = diameter of disc (mm)

The width of cut of disc harrows measured and worked out from the above formulae should be compared and reported in the test report.

ii) Effective field capacity

The actual output in terms of area covered per hour is expressed as the effective field capacity. In calculating the effective field capacity, the time consumed for real work and that lost for other activities such as turning, adjustments etc. should be taken. Time for refueling should be deleted because usually filling up before starting the test can make refueling unnecessary except for specially large field. It can be calculated as :

$$E_e = \frac{A}{T_p + T_N}$$

Ee = Effective field capacity (ha/h)
A = Area covered (ha)
T_p = Productive time (h)
T_N = Non productive time (h)

(Non productive time is the time lost for turning and adjustments etc. excluding refueling and machine trouble)

iii) Field efficiency

The field efficiency is the ratio of effective field capacity to the theoretical field capacity expressed as percentage.

$$\text{Theoretical field capacity} = \frac{\text{Theoretical width of implement(cm)} \times \text{speed of operation(m/sec)} \times 36}{10000}$$

$$\text{Field efficiency}(\eta) = \frac{\text{Effective field capacity (Ee)}}{\text{Theoretical field capacity (Et)}} \times 100$$

For most of the tillage operations field efficiency ranges between 75-90%.

b. Quality of work

i) Depth of cut

The vertical distance between furrow sole and ground level is referred as depth of cut. To obtain accurate result, the depth should be measured at minimum 10 places and its average taken .

ii) Soil inversion

The soil inversion is the process through which the furrow slice is inverted. The inversion characteristics can be measured by weed count method. In this method the number of weeds/stubbles present are counted before and after ploughing with the help of placing a square ring of 30 cm x 30 cm or 50 cm x 50 cm at random in the field and at minimum 5 locations. The soil inversion is expressed on percentage basis.

$$\text{Soil inversion (\%)} = \frac{WB - WA}{WB} \times 100$$

Where: WB = No.of weeds present per unit area before the operation.

WA = No. of weeds present per unit area after the operation.

iii) Soil Pulverization

Soil pulverization is the process of breaking of soil into small aggregates resulting from the action of tillage forces. The mean mass diameter (NMD) of the soil aggregated is considered as index of soil pulverization and can be determined by the sieve analysis of the soil sample through a set of standard test sieves (IS: 460-1982). Sieving provides a simple means for measuring the range of clod size and relative amount of soil in each size class.

Procedure

i. Collect the material from an area of 150 x 150 mm and to the depth of operation of the implement.

ii. Weigh the material accurately and pass it through a set of sieves with aperture size 11.2, 8.0, 5.6, 4.0, 2.8 and 2.0 mm.

iii. Weigh the material shifted through each sieve and the material retained on 11.2 mm sieve.

iv. Calculate the percentage of mass of material shifted through each sieve and that shifted and retained on 11.2 mm sieve. The material passed through and retained on 11.2 mm sieve should be taken as cent per cent.

v. Plot a curve (See Fig.4.3) with sieve size and percentage of mass material shifted through each sieve.

vi. Obtain area of the shaded portion of the curve by planimeter.

vii. Obtain MMD by multiplying the area of the shaded portion with MMD equivalent to one unit area in the graph.

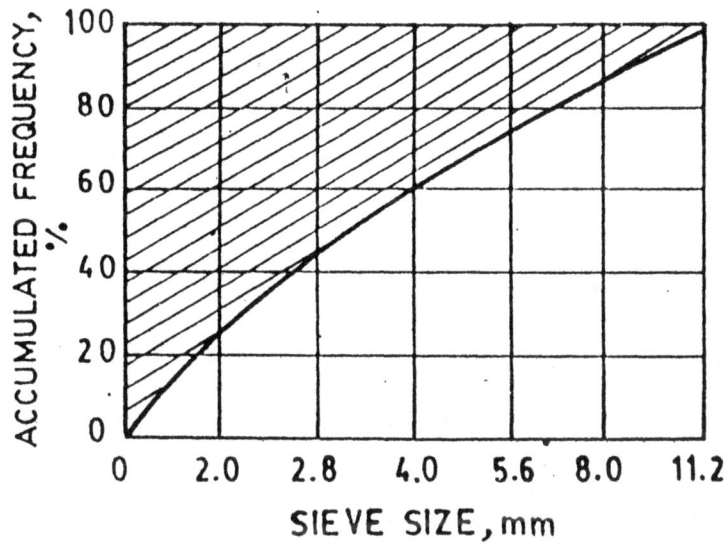

FIG.4.3 SIEVE ANALYSES CURVE FOR MEAN MASS DIAMETER

Another method

Pass the soil sample through a set of sieves. Weigh the soil retained on the largest aperture sieve, passed through each sieve and retained on the next sieve and passed through the smallest aperture sieve. Mean mass diameter can be calculated as shown in Table 1.

Table 1: Mean mass diameter calculation method

Size of aperture (mm)	Dia. of soil passing the upper sieve and retained on the next small aperture sieve (mm)	Representative dia. of soil (mm)	Weight of soil (Kg)
2	<2	1	A
2.8	2-2.8	2.4	B
4.0	2.8-4.0	3.4	C
5.6	4.0-5.6	4.8	D
8.0	5.6-8.0	6.8	E
11.2	11.2 >		F

$$MMD = \frac{1}{W}(A + 2.4B + 3.4C + 4.8D + 6.8E + xF)$$

Where : MMD = Mean mass diameter

W = A+B+C+D+E+F

x = Mean of measured dia of soil clods retained on the largest aperture sieve.

C. Draft measurement

i) For manually operated tools

Since ready made dynamometers for manually operated tools might not be available at present, special measuring apparatus must be devised and fabricated. Although, recording or indicating type strain gauge meter can be utilized as a part of the apparatus, but is expensive and too sensitive to field conditions. Simple draft measuring devices for manually operated tools have been developed by Central Institute of Agricultural Engineering, Bhopal.

ii) For trailed type implement

The draft required for pulling the implement can be measured by inserting a spring/ hydraulic or strain gauge type dynamometer between the hitch of the implement and power source. If the line of pull through the dynamometer is not horizontal, the angle of inclination with horizontal plane should be recorded and calculated as follows:

$$D = P \cos \theta$$

Where :

D = Draft (kgf)

P = Pull measured by a dynamometer (kgf)

θ = Angle between the line of pull and the horizontal (deg).

iii) For mounted type implement

Draft can be measured by two tractors and a dynamometer. An implement is mounted on a tractor (A) and this tractor is towed by another tractor (B) through a dynamometer as shown in Fig.4.4. Measure the draft when the tractor A is in neutral gear condition but implement in operating condition and again when the implement is in lifted position. The difference between two readings gives the draft requirement of implement.

Draft power requirement for the operation of implement can be calculated by the following formula :

$$Power (PS) = \frac{Draft (Kgf) \times Operational\ speed\ (m/sec)}{75}$$

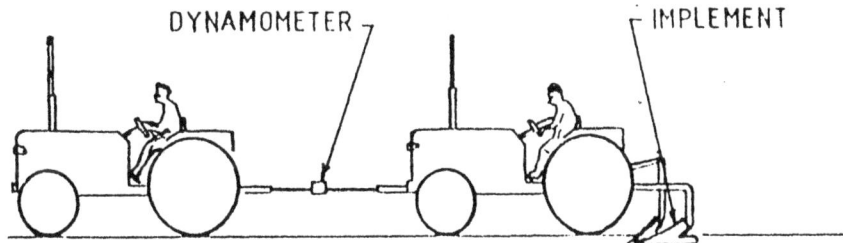

FIG. 4.4 MEASUREMENT OF DRAFT FOR
MOUNTED-TYPE IMPLEMENT

D. Fuel consumption

The tank is filled to full capacity before and after the test. Amount of refueling after the test is the fuel consumption for the test. While filling up the tank, careful attention should be paid to keep the tank horizontal and not to leave empty space in the tank. The fuel consumption will give an idea of energy requirement by the implement for the operation.

E. Soundness of construction

During entire period of testing a complete record of the defects and breakdowns should be recorded and reported in the test report. These would reflect to a great extent the soundness of construction of the implement under test.

F. Ease of adjustments and operation

A complete record of provisions for various adjustments to cover a wide range of operating conditions and ease of carrying out the adjustments, should be maintained during the test. Apart from this, the ease of operation of the implement should be observed critically and reported.

4.5. Interpretation of test results

Although evaluation of implements on the basis of their performance tests is a complex job because of number of variables affecting the performance of the implement directly or indirectly during operation. These are soil type, moisture content, presence of weeds/stubbles, power source, provision of adjustments and skill of operator, etc. It is difficult to maintain identical field conditions to facilitate comparison of field results. However, careful evaluation of the test report can give realistic information. provided, the testing has been conducted following standard code and procedure. An implement requiring the minimum draft for operation is considered to be the best, provided, the quality of work done is also satisfactory. In case of tractor/power tiller operated implements, the maximum fuel economy will be achieved at minimum draft requirement. For determining the fuel economy, comparison of fuel consumption per unit volume of soil worked will provide a realistic data to evaluate the implement. Lesser the fuel consumption per unit volume of soil worked, better is the quality of implement.

The rate of work is a function of size of implement, speed of operation, size of plot and field parameters. Thus, the evaluation of implement on the basis of their rate of work is done by comparing the result with performance results of similar implements. However, an implement giving higher output without sacrificing the quality of work is considered better.

Soundness of construction, ease of operation and adjustment and rate of wear of critical components of the implements are also important factors to be given due weightage in their evaluation. An implement of a simple design and robust construction should be preferred over sophisticated implement having weak construction. Operating cost of the implement is another important point which is to be considered in evaluation of implement. However, quality of work cannot be sacrificed for low operational cost. The availability of efficient after sale service facilities including warranty, cost and availability of spares, use of standard size hardware and other fast moving components are also important considerations.

During the process of evaluation it may be possible that the implement may not have all the plus points as described in testing procedure. But it is expected that test results should conform to the relevant Indian Standard, quality of work done is satisfactory, economical in operation, sturdy construction, adequate provisions for adjustments to cover a wide range of working conditions and safety aspects have been provided in the implement. Moreover, an overall performance index be calculated on the basis of various performance results obtained during testing and compared with other similar products.

Chapter 5

TESTING AND EVALUATION OF SEED-CUM-FERTILIZER DRILLS

5.1. Introduction

Seed drills and seed-cum-fertilizer drills of different types and capacities are now being extensively used in the country for sowing different kinds of seeds and placement of fertilizer. Placement of seeds at correct depth is very important for proper germination of the seed especially under dryland farming where soil moisture is at greater depth. Similarly the placement of fertilizer is also very important for maximum fertilizer utilization efficiency. The fertilizer should not come in contact with the seed to avoid chemical injury. Therefore, precision placement of seeds and fertilizer is necessary for achieving perfect standing of crop. It also reduces sowing time and thus overcomes the shortages of labour.

5.2 Functional requirements of seed-cum-fertilizer drill

1. It opens furrow to a uniform depth and spacing.
2. It drops seeds and fertilizer uniformly without causing any injury to seeds.
3. It drops seeds and fertilizer at pre set rates.
4. It maintains proper distance and depth between seeds and fertilizer.
5. It covers the seeds and compacts the soil around them in order to conserve moisture.

5.3. Major components

Out of all the designs of the seed-cum-fertilizer drills used in the country at present, the one employing external flutted-feed rollers for seed metering is popular and widely used. The major components of such a drill (Fig.5.1) are frame, seed hopper, fertilizer hopper, furrow openers, transport wheels, depth adjuster, seed metering mechanism, fertilizer metering mechanism, drive wheel, sprocket and chain mechanism, seed rate indicator, seed tubes, fertilizer tubes, tynes and hitching links.

5.4 Critical components

Generally following components require replacement after they get damaged or wornout :

1. Seed metering mechanism
2. Fertilizer metering mechanism
3. Furrow opener
4. Seed and fertilizer tube
5. Sprocket and chain
6. Hitch pin

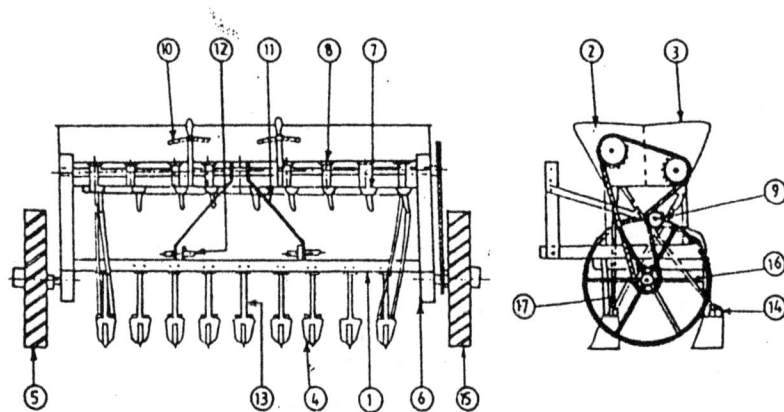

1. FRAME   2. SEED HOPPER   3. FERTILIZER HOPPER   4. FURROW OPENER   5 TRANSPORT WHEEL
6. DEPTH ADJUSTER   7 SEED METERING MECHANISM   8 FERTILIZER METERING MECHANISM   9. SPROCKET
CHAIN TIGHTNER   10 INDICATOR   11. HITCHING LINK   12. HITCH PIN   13. TINE   14. SEED BOOT
15. DRIVING WHEEL   16. SEED TUBE   17. FERTILIZER TUBE

FIG. 5.1 MAJOR COMPONENTS OF SEED-CUM-FERTILIZER DRILL

5.5 Type of tests

The following aspect of the machine's performance shall be assessed:

A. Laboratory tests

1. Specification checking
2. Stationary calibration
3. Testing on repeatability of notch mark setting for calibration.
4. Effect on seed discharge rate due to depth of seeds in the seed hopper
5. Effect on seed discharge rate due to different forward speeds .
6. Evenness of seed spacing in the row.
7. Seed specifications
8. Seed damage
9. Chemical analysis and hardness test of critical components

B. Field test

1. Field calibration at different settings
2. Rate of work
3. Relative placement of seeds and fertilizer
4. Quality of work

5. Power requirement
6. Fuel consumption
7. Labour requirement and field efficiency
8. Ease of setting, operation and adjustment
9. Soundness of construction ·

5.6 General conditions

i. Selection of machine

The test sample shall be selected at random from the production line by a representative of the testing station. The method of selection shall be specified in the test report. The test machine should strictly conform to the description and specifications submitted by the applicant and as generally offered for sale, except when the machine is prototype and is not on mass production and submitted for confidential test.

ii. Running-in and preliminary adjustments

The machine should be fully assembled, examined in detail and run-in as per recommendations of the manufacturer, before the commencement of the test. At this stage, the applicant is encouraged to depute his representative for necessary adjustments and to demonstrate the operation of the machine to the testing staff. The test team should preliminarily run the machine to familiarize themselves with the operation and to check clarity of the instruction booklet. All the adjustments made, should be in accordance with the instructions contained in the Operator's Manual/literature/other written instructions on adjustments issued by the manufacturer. Adjustments made during test period which are not in conformity with the manufacturer's recommendations, shall be reported in the test report.

iii. Servicing and maintenance

The machine shall be serviced as per manufacturer's recommendations. All the tests shall be done with the seed box three-forth full. However, if necessary, this shall be repeated with half and full seed box.

5.7 Test conditions

A. Test conditions of seeds and fertilizer

a) Seeds

    i) Scientific and popular name and variety
    ii) Germination rate in the laboratory. Germination rate is measured by counting the number of germinated seeds in a laboratory dish in which filter paper is laid and water supplied. Usually, the laboratory dishes are put into

an incubator at 25-30°C. Ratio of germinated seeds to all seeds after three days is called germination rate.

iii) Bulk density

It is measured by filling the material in the cube or cylinder of which the volume and weight is known.

iv) Average size

It is measured to distinguish size between different types of seed and varieties.

b) Granular fertilizer

     i) Name and kind of fertilizer
     ii)Distribution of granule size

It is measured by sieving (1.0, 2.0, 2.8, 4.0 mm sieves) using a mechanical sieve shaker. Sieving time shall conform to the instructions provided by the sieve manufacturer.

     iii)Bulk density
     iv)Angle of repose

There is a relation between the angle of repose and the discharge rate. Therefore, it is measured and relationship between angle of repose and discharge rate is established.

v) Moisture content

Moisture content of granular fertilizer is measured by drying 5 hours at 100 °C. In case of urea containing fertilizers the drying period should be 4 hours at 75°C.

B. Test Condition of machine

     a- Source of power : manual, animal drawn or tractor drawn
     b- Adjustment of parts: metering shaft speed, delivery opening.adjustment, row spacing etc.
     c- Test condition of field :
          i) Area and shape of test plot
          ii) Type of soil
     d- Method of land preparation and size distribution of soil clods at surface layer (under 1cm, 1-2cm, 2-3cm, 3-4 cm over 4 cm.)
     e- Soil moisture content and bulk density
     f- Atmospheric conditions.

C. Test Condition of operation

      a- Setting of seed and fertilizer rate
      b- Setting depth of drilling

5.8. Procedure for laboratory testing

1. Specification checking

The manufacturer shall supply all specifications about the machine in the prescribed form. These shall be verified by the Testing Station. Any discrepancy if noticed, shall be mentioned in the test report.

2. Stationary calibration

For the laboratory calibration, three different type of seeds shall be selected from the applicant's recommendations. If no recommendation has been made by the applicant, the common seeds used in the area should be selected.

    For laboratory calibration, the machine is tested on the special seed drill testing rig which can be fabricated at the testing station. The drive to the feed shafts is given in the same ratio as (Fig.5.2) in the actual machine by an electric motor.

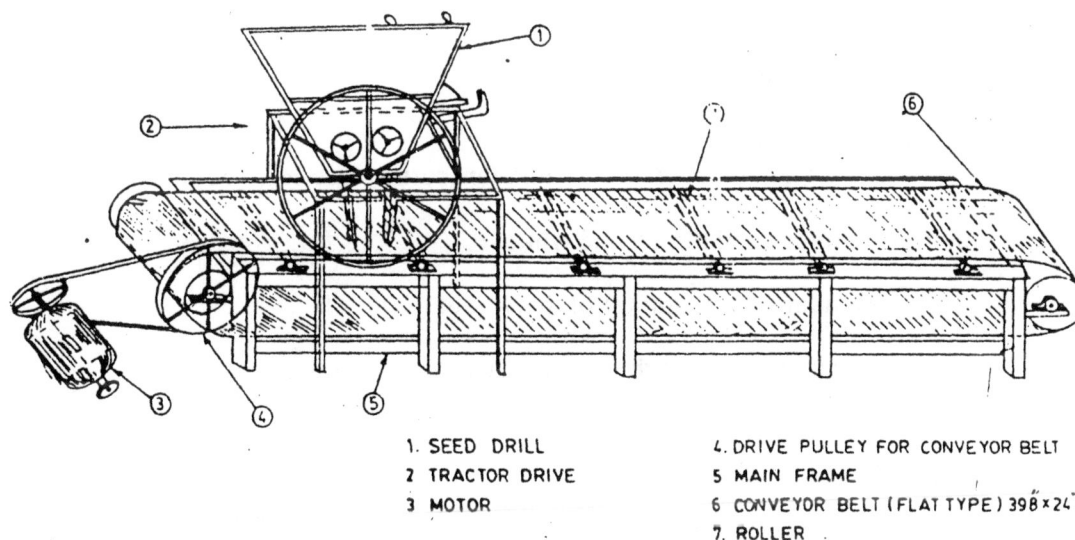

1. SEED DRILL	4. DRIVE PULLEY FOR CONVEYOR BELT
2 TRACTOR DRIVE	5 MAIN FRAME
3 MOTOR	6 CONVEYOR BELT (FLAT TYPE) 398×24
	7. ROLLER

FIG.5.2 SCHEMATIC DIAGRAM FOR SEED DR.LL TEST RIG

Calibration of machine

Laboratory calibration is required to check correctness of seed and fertilizer dropping rates. The procedure is as follows:

Material required for calibration

a- About 40 kilogram of seeds or the seeds required for sowing one fourth of hectare. The variety of the seed should be same as to be sown.
b- Weights and balance to weigh the seeds
c- Tape of 25 meter
d- A cloth piece to tie at one place on the drive wheel of the drill.
e- Cotton cloth bags or plastic bags to collect the seeds and fertilizers coming out from each of the seed/fertilizer cup.
f- Two jacks or wooden blocks for jacking up the machine above the ground, so that drive wheels could be rotated freely.

Procedure for calibration

i-Measure the distance between two adjacent furrow openers. Check inter furrow distance and if necessary adjust to the recommended value. Count the number of furrow openers.

ii-Effective working width of the drill:

$$W = \frac{N \times d}{100} \text{ meter}$$

Where N = number of furrow openers
d = distance between two adjacent furrow openers (centimeters)

iii- Circumference of the driving wheel (L):

Measure the diameter of the wheel (cm) = D

Circumference of the wheel (L) = $\frac{22 \times D}{7 \times 100}$ meters

iv- Area sown in one revolution of the drill (A):

A = W x L square meter

v- Number of revolutions required to sow one hectare (R):

$$R = \frac{10,000}{A} \text{ revolutions}$$

vi- Number of revolutions actually required to cover one hectare, considering 10% wheel slip (M) :

$$M = \frac{R\ (100\text{-}10)}{100} \quad \text{revolutions}$$

vii- Seed rate to be sown per hectare = P Kg (say)

viii- Seed quantity required for one revolution(s):

$$S = \frac{\text{Seed rate (kg)}}{\text{No.of revolutions required to sow one hectare}}$$

$$= \frac{P \times 1000}{M} = G\ gm\ (say)$$

ix- Seed quantity required for 50 revolutions:

$$G \times 50 = B\ grams\ (say)$$

x- Similarly calculate fertilizer quantity required for 50 revolutions (say 'C' grams)

xi- After the above calculations, prepare the drill for calibration in the following manner.

a) Jack up the seed drill:
Remove the tubes from the feed cup and in their place tie plastic or cotton bags to collect the seed and fertilizer. The bags should be numbered.

b) Tie a piece of white cloth on drive wheel to count the number of revolutions.

c) Rotate the drive wheel 50 times at normal field speed.

d) Weigh the seeds and fertilizer in each bag. The inter furrow variation should not exceed ± 7%. If the variation is higher, the drill should be thoroughly checked and defect rectified. The calibration should be repeated. If the variation continues, the same should be reported in the test report.

e) The seeds of all the bags be collected in one big bag and weigh it. Similarly weigh the fertilizer also. These quantities should be equal in weight as calculated above at point ix and x. If this is less or more, the metering mechanism should be adjusted accordingly and calibration process be repeated till we get desired seed and fertilizer rate. A firm mark should be put on the indicator setting accordingly.

Field Checking

Many owners do not calibrate their seed drills but check the operation in the field. This is done as follows :

    a- Set the drill at the desired seed rate through indicator.
    b- Set the drill in the field where an area of about 1/25 hectare has been marked for sowing.
    c- Fill the seed box or hopper with seed upto a mark, or to the top and level off carefully.
    d- Sow exactly 1/25 hectare as marked the field.
    e- Refill the seed box from a bag of which exact weight is known.
    f- By reweighing the grain left in the bag and subtracting from the original weight of the bag, will indicate quantity sown. If the quantity is not correct make compensation and adjust in the seed drill and check again. The field checking method is not recommended for laboratory test.

3. Testing on repeatability of notch-mark setting for calibration

The machine is calibrated (following stationary calibration method) at three random positions on the indicated scale with three replications. Each replication is attempted after shifting the calibration lever several times. The objective is to ascertain the accuracy of the notch setting in terms of repeated operations.

4. Effect on seed discharge rate due to depth of seed in seed hopper

The effect of seed depth in hopper on seed discharge rate is determined keeping the hopper full to 3/4 full, 3/4 full to 1/2 full and 1/2 full to 1/4 full. The other variations such as position of seed drill, position of seed adjusting lever shall be kept constant. The variation in seed discharge due to box filling should not exceed 10%.

5. Effect on seed discharge rate due to different forward speeds

The difference in seed discharge rate would be assessed by using recommended seeds and variation in discharge at three different speeds shall be recorded. The variation in seed discharge due to different speed should not be more than 15%.

6. Evenness of seed spacing in the row

A sticky belt is passed under the machine at equivalent operating speed as in the field and drive to the feed shaft is also given corresponding to the same speed. The pattern of seed falling on the belt is recorded. The mean spacing shall be recorded and suitable index of variation from mean will be used to express the uniformity of spacing. Three replicate runs shall be carried out and results of each run shall be recorded.

7. Seed specifications

The bulk density and number of seeds in one kilogram shall be determined for all types and varieties of seeds to be used during testing before actual testing of seed drill is carried out.

8. Seed damage test

The test is conducted to see the visual damage caused to the seeds passing through the metering device. Grain samples from the seed hopper as well as from the seeds passing through the metering device are taken separately and analyzed for external and internal damage and reported.

9. Chemical composition and hardness test of critical components

The chemical composition and hardness of furrow opening device and other critical components as decided by the testing authority shall be determined and reported. The weight of furrow openers at the start and completion of test shall be used to determine rate of wear of furrow openers.

5.9. Procedure for field testing

1. Field calibration at different settings

Field calibration is done with the same type of seeds which are used for laboratory calibration. A 100 meter run of well prepared seedbed is used for field calibration. The condition of the seedbed is recorded. The machine is run for actual sowing operation except that the seeds dropped from the different spouts are collected separately and weighed. The number of revolutions of the ground wheel are also recorded to determine the wheel speed effect on seed rates in comparison to that obtained in the laboratory. The calibration is done at the same notch settings as during laboratory test and reported accordingly.

2. Rate of work

The machine is run continuously in the field for about 25 hours under different field sizes and conditions to assess the working capacity of the machine and to record any breakdowns or excessive wear. The net rate of work excludes the turning time at head land, time for filling seeds in hopper and other stoppages.

3. Relative placement of seeds and fertilizer

Operate the drill in the field under the normal seedbed conditions and with average depth setting of the furrow openers. Cover at least 100 meter length. Then carefully remove the soil without disturbing the seeds and the fertilizer at minimum 5 different places in each row. Measure the depth of the seeds below the soil surface and the vertical spacing of the fertilizer with respect to the seeds. Measure the horizontal spacing also. It will be much easier to locate the seed and fertilizer if maximum rate of application are used. The fertilizer is more easily located if white lime is mixed with it. Repeat the procedure at minimum and maximum depth setting of the furrow openers. Average value of relative placement be reported.

4. Quality of work

The quality of work including coverage of seeds, row spacing, blockages and wheel skids would be recorded during field work.

5. Power requirement

The draft required to move the machine in a normal seedbed is observed by using an appropriate hydraulic dynamometer. The draft is measured at maximum, minimum and average depth of sowing and with 3/4th full seed hopper. Detail procedure of power measurement for trailed and mounted type seed drill is given below:

a- Draft measurement

i) For trailed drill

Insert a hydraulic dynamometer in the hitching point to measure the draft. The draft is defined as the horizontal component of the pull, parallel to the line of motion. If the line of pull through the dynamometer is not horizontal, then measure the angle between the line of pull and the horizontal line. Calculate the draft of horizontal component. Mark a 20m run space in the middle of a long row in the field with an easily distinguished poles. Start the drill well in advance of the first pole and be sure it is operating smoothly when it reaches this pole. As the drill travels the marked run-length, record the dynamometer reading. The more he readings taken the better the results are. Calculate the average of all the readings taken within a particular run. At the same time, record the time, the machine took to cover the marked run length. From this value calculate the speed of travel in metre per second. Also calculate the wheel slip and theoretical field capacity. The power can be calculated as follows:

$$\text{Metric horsepower (Ps)} = \frac{\text{Draft (kgf) x speed (m/s)}}{75}$$

ii) For mounted drill

Mark a 20m run in the field with the help of poles. Measure pull of the tractor to be used for seed drill operation by pulling it with another tractor having dynamometer in between both tractors. Similarly pull of same tractor with seed drill in operation is noted in the same marked run. The draft of seed drill is computed by difference of these two pull readings. Time taken to cover 20 m is also recorded to calculate speed of operation. The power requirement is calculated with the help of above mentioned formula.

6. Fuel consumption

The fuel tank of tractor is filled to its full capacity before and after test. The quantity of fuel filled at the end of test divided by total hours of operation will give hourly fuel consumption.

7. Labour requirement and field efficiency

The actual labour requirement for operating seed drill in the field should be assessed and recorded. The theoretical field capacity is the rate of field coverage that would be obtained if the drill was operating continuously without interrupting like turning at the ends, filling of hoppers, etc. The effective field capacity is the actual rate of work which includes the time lost in filling hoppers, turning at the end of rows, cleaning of openers, making adjustments etc. The theoretical field capacity can be determined while taking data for draft requirement. The effective field capacity is the actual area covered during test. From these values field efficiency is computed.

8. Ease of setting, operation and adjustments

During the field operation of the machine, all the operational and adjustment difficulties are recorded to assess the handling characteristics.

9. Soundness of construction

All major and minor breakdowns/damage of the parts occurred during entire test period are recorded and reported in the report.

Field test results can be observed and reported as per performa shown in Appendix-VI (a) & VI (b).

Chapter 6

TESTING AND EVALUATION OF RICE TRANSPLANTER

6.1 Introduction

The importance of rice cultivation in India has considerably increased with the introduction of high yielding varieties which have made rice more paying than other Kharif crops. Paddy occupies maximum area under cultivation in India. The operations like transplanting, interculture, harvesting and threshing etc. are very tiresome and labour consuming. The scarcity of labour poses a big problem to complete the operations particularly transplanting in time. The transplanting period of high yielding varieties is very short and limited. The delayed translating causes progressive decrease in the yield and in Northern region very late transplanting may result in complete failure of crop due to cold weather at the stage of maturity. Therefore, rice transplanter is an important machine which can help the farmers in completing the work in time and achieve desired quality of work.

6.2 Classification of rice transplanters

A-Based upon power source

1- Manually operated
2- Power tiller operated
3- Self-propelled type i) Walking type
ii) Riding type

B-Based upon type of seedlings

1- Root-washed seedlings
2- Mat type seedlings

C-Based upon travelling type

a- Walking type
b- Riding type

D-Based upon transplanting mechanism

a- Crank mechanism
b- Rotary mechanism

6.3 Nursery for transplanter

Traditionally grown nursery in India is suitable for root washed type transplanters.

However, transplanter using mat type nursery have been widely adopted in Japan, where this operation is completely mechanized. For raising mat type seedlings, trays or frames placed on plastic sheet are required. Soil is filled in these frames/trays which have generally 1-8 cm height. In Japan, mat type nursery is grown on a commercial level. In India, successful efforts have been made to raise mat type nursery in field. Labour requirements at the time of sowing are comparatively larger while for nursery uprooting is lesser in mat type nursery.

Mat type nursery is, generally transplanted in the age group of 25-40 days. At this age, the leaf stage is minimum 2 and the minimum plant height is 12.5 cm. Nursery mats after uprooting are directly placed on the machine for transplanting.

6.4 Construction of rice transplanter

Rice transplanter consists of a prime mover, wheels, float, transmission system, planting mechanism, seedling feeding mechanism and operation control device etc. The detail of construction for walking type and riding type transplanter is shown in Fig.6.1 & 6.2. The traveling speed is usually 0.3-0.7 m/s and in case of high performance transplanter speed could be 1 m/s.

The vertical distance between soil surface and hard pan is not constant in the field. Thus depth of planting depends upon change in the relative position of the float and planting finger. Accordingly height relation between wheels and floats influence planting accuracy. Generally, transplanters have automatic planting depth control to maintain uniformity.

Power transmission system is meant for transmitting power from prime mover to planting device and traveling device. Changing ratio of planting speed and traveling speed can make adjustment of hill spacing. The system consists of a set of gears, belts, roller chain, main clutch, safety clutch etc.

Planting mechanism is either crank type or rotary type. Generally walking type rice transplanter are equipped with crank type mechanism and riding type with crank type or rotary type mechanism. Crank type mechanism consists of four bar linkages (crank, rocker, coupler and fixed link), planting arm, planting finger and planting fork. This mechanism makes a unique motion on a locus of the point of planting finger according to the rotating crank. Rotary type mechanism consists of eccentric planetary gears, planting arm, planting fingers and planting forks. This mechanism enables the high speed planting work comparing with crank type.

Seedling feeding mechanism consists of cross-feeding and longitudinal feeding mechanism. The cross feeding mechanism moves mat seedlings sideways by the reciprocation of seedling trays and the longitudinal mechanism slides down the mat seedlings on the seedling trays.

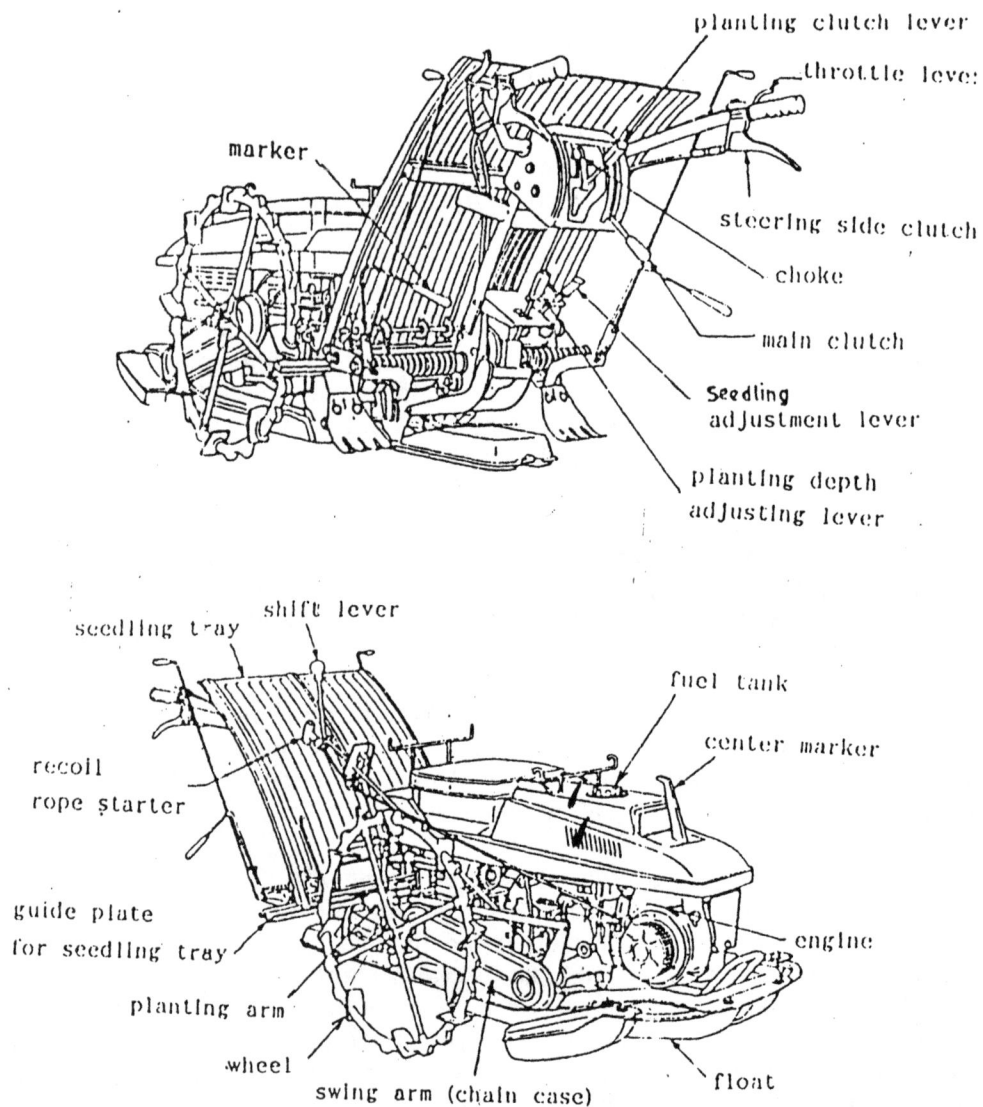

FIG. 6.1 CONSTRUCTION OF WALKING TYPE RICE TRANSPLANTER

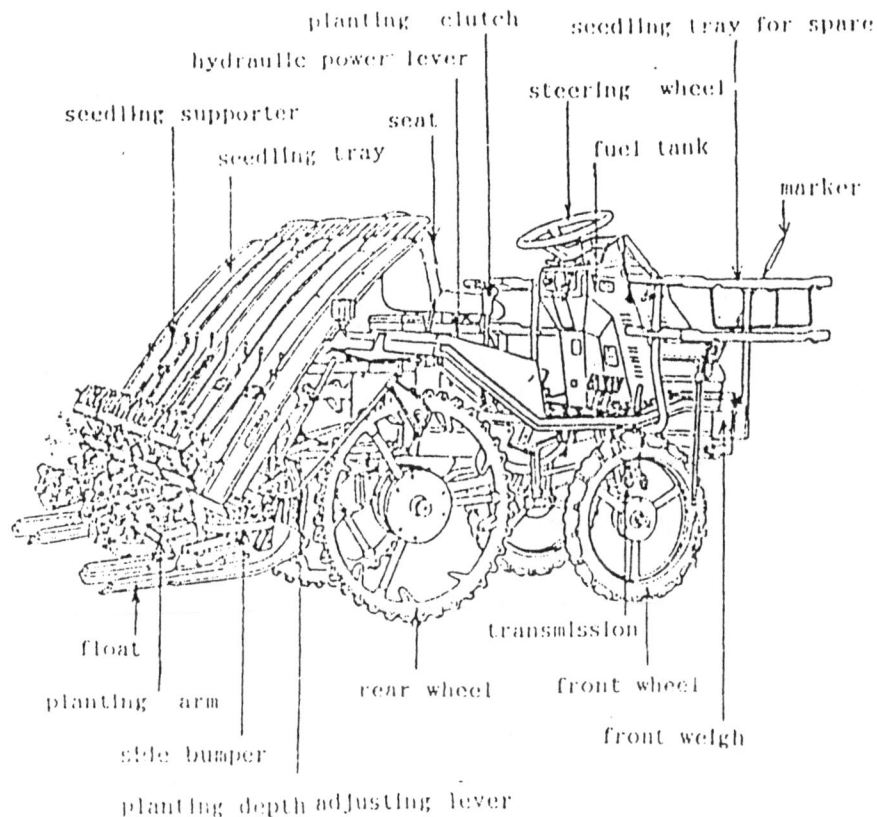

FIG.6.2 CONSTRUCTION OF RIDING TYPE RICE TRANSPLANTER

6.5 Scope of test

The test procedure described below is formulated to assess the performance of transplanter, mainly based upon field test which is represented by work accuracy and work efficiency. Work accuracy in other words is planting accuracy which means the reliability of transplanting operation and work efficiency means working capacity. On field performance test, field and seedling conditions are very important, as these will effect work performance. For example, if short seedlings are used in test field, having very soft soil surface, then work accuracy will be poor due to high percentage of buried hills. Therefore, testing of rice transplanter should be carried out in normal conditions and that too should be defined in the test report.

6.6 Terminology

a) Root-washed seedling

These seedlings are those which are uprooted from the traditional nursery field. After uprooting, the soil attached to the roots is washed away. Then, seedlings are separated from each other and the roots are cut to a length of 2-3 cm, so that higher accuracy of planting operation can be achieved.

b) Mat-type seedlings

These seedlings are raised in seedling boxes or trays. Seedlings grown in trays get their roots entangled forming a root-mat. The root-mat, including the seedling, is taken out and placed on the transplanter.

c) Leaf stage of seedlings

Leaf stage indicates the character and length of the seedlings. Normally seedlings are transplanted by machine at the stage of having not less than 2 leaves and not more than 6 leaves.

d) Density of seedlings

In case of mat type nursery, the number of seedlings obtained in a section of 2 cm by 2 cm shall be converted into number of seedlings per unit area (cm2). It influences the accuracy of planting operation and number of seedlings per hill.

e) Rate of work

The area covered in the field per unit time.

f) Field efficiency

During the field operation, time will be lost in turning at the headland, corners and other defects. These affect the efficiency and decreasing the rate of theoretical field capacity of work. Moreover, field efficiency will vary according to the size and shape of the field, the type and size of machine, the skill of the operator and other similar factors.

g) Slippage

The slippage of driving wheels is calculated by the following formula:

$$\text{Slippage (\%)} = \frac{N_2 - N_1}{N_2} \times 100$$

Where N_1 : The number of revolutions of driving wheels for a certain distance in the puddled field.

N_2 : The number of revolutions of driving wheels for a certain distance on the hard surface

h) Damaged seedlings

These can be divided into two categories. Damage is caused by cutting or bending of the seedlings and by internal damage of the growing point of the seedling due to crushing by planting fork.

i) Floating seedlings

During transplanting, the roots get into soil and seedlings remain settled for some time. But some times due to higher water level and disturbance by transplanter, seedlings get up-rooted and float on the water. Such seedlings are called floated seedlings resulting in missing hill.

j) Buried hills

Hills which are buried under the soil after transplanting due to movement of soil in the already transplanted rows, caused by machine travel are called buried hills.

k) Missed transplantings

It denotes the rate of missing hill. It is expressed in percent.

l) Total missing hills

Total missing hills (%) is the sum of percentage of buried hills, floated hills and missing hills caused by planting mechanism and unevenness of seedling density in mat.

6.7 General conditions

1- Condition of rice transplanter

The test sample shall be selected randomly from the production line and shall be well run in.

2- Preliminary trial

The rice-transplanter shall be tried and adjusted in the puddled field before field performance test.

3- Fuels and lubricants

Fuel and lubricants used for test shall conform to specification prevailing in the country and be easily available in the market.

4- Seedlings

Seedlings used shall be selected by testing authority. Two kind of seedlings preferably 2-4 leaves shall be used during testing.

5- Field dimensions

The test plot should have length of 50 m and width 25m.

6- Measuring instruments

All measuring instruments shall be inspected and calibrated before use.

6.8 Test conditions

As machine performance and operating accuracy will vary considerably according to seedling conditions. character of soil. water depth at the time of planting and adjustment of working parts of the machine. test conditions have to defined as shown below:

1- Field conditions

    a- Area and shape of test field
    b- Type and character of soil
    c- Last crop grown in the field
    d- Application of organic matter, if any
    e- Method of land preparation
    f- Period after puddling
    g- Hardpan depth: The depth from soil surface to hardpan. This is measured by the length of a thin stick penetrated into soil.
    h- Water depth: The depth of water over the soil surface.

2- Seedling conditions

    a- Variety of rice
    b- Type of nursery
    c- Soil type of seedbed or nursery field
    d- Date of sowing
    e- Quantity of seed
    f- Germination rate

g- Nursery duration

h- Plant height: The length of a plant from the root base to the top of leaf.

i- Leaf stage: The number of leaves except a coleoptile.

j- Plant establishment density in mat:- The number of plant per unit area (cm2).

3- Setting conditions

a- Number of workers
b- Traveling speed
c- Number of seedlings per hill
d- Row and hill spacing
e- Position of each adjusting parts
f- Any other item

6.9 Test items

1- Specification checking
2- Field performance test

a) Work accuracy

i - Total missing hills
ii - No.of seedlings per hill.
ii - Planting depth
iv - Row spacing
v - Hill Spacing
b) Traveling speed :
c) Working time
d) Actual planted area
e) Rate of work
f) No. of mats used
g) Slippage
h) Time lost in turning and other stoppages
i) Field efficiency
j) Draft
k) Fuel consumption
l) Labour requirement
m) Turning space
n) Turning radius

A suggested performa for field observation is shown in Appendix-VI (c).

3- Handling test

The object of this test is to ascertain the easiness of handling and adaptability for the rice transplanter. The testing authority shall investigate the adjustment of each mounting parts and measure vibration, noise and so on. Moreover, if necessary measure and investigate the work performance under the different field conditions.

4- Water proof test

The object of the test is to ascertain dust and water proof function of mainly wheel axles and planting portion and to ascertain any abnormality or trouble in any of the parts. This test should be conducted in a water-tank to simulate the field conditions. The transplanter should be operated under the operating conditions as for as possible w.r.t. important parameters like depth of planting finger. The period of testing time should be 15 hours.

5- Investigation after disassembling

After all test are completed, the testing authority shall disassemble and check the rice transplanter. The object of this test shall be to check the abnormality of critical components/assemblies and reported accordingly.

6.10 Some suggested limits of rice transplanter

a- No.of plants per hill - 3 to 5 plants
b- Planting depth - 2 to 3 cm
c- Row space - 20 cm
d- Hill spacing - 10-15 cm.
e- Missing hill (%) - less than 10

Chapter 7

TESTING AND EVALUATION OF IRRIGATION PUMPS

7.1 Introduction

Centrifugal pump is one of the most important water lifting devices for irrigation purpose. These are very popular because of low initial cost, greater flexibility in application, wide working range in size and capacity, simplicity in construction, constant steady discharge and ease of operation, maintenance and repairs etc.

In the last decade, there has been a four-fold increase with consumption of energy in the agricultural sector which accounts for about 17% of the total energy consumption in the country. Of this, roughly 85% energy is consumed by the agricultural pumping systems.

It has been reported that about 56 lakhs diesel and 64 lakhs electric driven pumps are being used by the farmers to-day for irrigation purposes. It is estimated that about ten lakhs pumping systems are added to this huge number every year. Some of the pumps used in agricultural activities are inefficient. The system efficiency is around 40%. Thus, there is collosal wastage of energy. If 90% of existing electric pumping systems are improved/rectified, 4000 to 5000 MW of power can be saved annually, i.e. a saving of more than eight thousand crores of Rupees/investment in power. It can also result to saving of about six hundred crores of Rupees annually in fuel alone. Therefore, measures will have to be taken to stop the tremendous energy drain generated by the new inefficient pumping systems. Some headway has been made by BIS (Bureau of Indian Standards) by formulating minimum performance standards for pumping system and related equipment. Norms specified in the standards need to be followed strictly and continuous effort made to improve system efficiency to an achievable level by following TQM (Total quality management)system in manufacturing of all the components as per ISO-9000 series standard formulated by International Standards Organization which has been adopted by BIS as IS-14000 series. The financial institutions and other departments concerned with agricultural subsidy and loans and the various marketing organizations should make it obligatory to deal in BIS marked equipments only.

There is an urgent need to launch a massive quality improvement programme to spread the concept of energy-efficient pump sets. For this, the operating range of different centrifugal pumps be specified for proper selection, as different pumps of the same size exhibit maximum efficiency at different total heads and discharge. Moreover, before a pump is marketed, it should be tested to make sure that it meets the guaranteed head, capacity and efficiency.

7.2 Construction of centrifugal pump

The centrifugal pump consists of following components:

i) Impeller: This is the heart of the pump and serves to exert a centrifugal force upon the water contained therein.

ii) Casing: This component performs the function of converting the centrifugal force created by the impeller effectively into pressure.

iii) Suction port: This is used to suck in the water.

iv) Discharge port: This is used to discharge the water.

7.3 Terminology

i) Suction lift : The vertical distance between the centre line of the pump shaft and the pumping water level in the well.

ii) Delivery head :The vertical distance between the centre line of the pump shaft and the centre line of delivery pipe at the discharge end.

iii) Static head: The vertical distance between pumping water level and discharge water level.
Static head = Static suction head+Static delivery head.

iv)Total head: The actual head against which the pump has to work
Total head = Static head + All head losses

v) Friction head: The head requires to overcome the resistance of water in fittings in the pipe line. It varies with (a) rate of flow (b) pipe size (c) interior condition of pipe and (d) type of pipe. This will include the friction in strainers, elbows, bends, footvalve, reducing sockets, tees, valves etc.

vi) Net positive suction head requirement (NPSHR): This is a function of pump design and varies from one make of pump to another, between different model of same manufacturerand speed of pump. Therefore NPSHR value must be obtained from manufacturer and for cavitation free performance. This value should be less than net positive suction head available.

vii) Velocity head: The vertical distance through which the liquid must fall to acquire a given velocity and is calculated using the following formula:

$$h = \frac{V^2}{2g}$$

Where h = velocity head
V = Velocity of water in the pipe
g = acceleration due to gravity

viii) Pump characteristics : The relation between speed, head, discharge and horse power of a pump are represented by number of curves known as "characteristics curve". Pump characteristics are represented by these curves.

ix) Specified speed : The speed of geometrically similar pump when delivering a volume of one cubic metre per second against a head of one metre height.

7.4 Selection of parameters

The performance of the total pumping system is the sum total of the efficiency of each part of the system. Besides, each part affects the efficiency/performance of the other part. It is, therefore, necessary that a "system approach is applied to the selection of a pumping system set.

Pumpset selection, is therefore, a process of selecting the prime mover, pump, footvalve/reflux valve, suction and delivery pipes, etc; keeping in view the type of soil. crops grown, area to be irrigated and the hydrogeological conditions, so that the pumpset gives timely and trouble-free services at the minimum cost per litre of water delivered.

For proper selection of pumpset, following parameters are applied:

1- Total head : It consists of the following :

a- Static suction head
b- Draw down
c- Delivery head
d- Frictional head

2- Discharge in litres per second desired from the pump. This will be determined by the following:

i) Size of the land to be irrigated.
ii) Crops to be grown
iii) Type of soil
iv) Capacity of the aquifer

For better results, from the pumpset, careful scrutiny of the following points is also necessary.

a- Depending upon the soil conditions, water losses due to seepage and evaporation should be considered at the time of calculating the water requirement.

b- Suction lift should be kept to the minimum possible and it should never be more than 6 metres.

c- One size higher of the suction, delivery pipe and foot valve/reflux valve should be used. This would result to fuel saving.

7.5 Causes for low overall efficiency of a pumping system

- Low efficiency of the pump
- Poor suction lift characteristics (NPSH) of the pump
- High friction losses in the foot valve/reflux valve
- Low efficiency of prime mover
- High specific fuel consumption in case of engine driven pump.
- Improper selection of the pump and prime mover such that they do not operate in their best efficiency zone.
- High friction losses in the pipes because of:
    a) High coefficient of friction of pipe material
    b) Under size pipes
    c) Sharp bends, unwanted lengths and heights of piping
- Air leakage in suction side
- Excessive power losses in power transmission from prime mover to pump
- Poor maintenance and lack of service facilities
- Improper installation.

Efficiency of pump

The efficiency has a direct bearing on power consumption. Higher the efficiency, more will be the energy conservation. An agricultural pumping set does not operate at a fixed duty point, like an industrial pump. When the pump starts, suction lift is low, but while pumping, the suction lift increases due to draw down. The total lift as well as total head increases. Again the operating points vary from season to season. Thus, agricultural pumps operate in a wide range of head and discharge. The efficiency of a pump should be reasonably good in the entire range and minimum declared efficiency should not be less than as shown in Fig.7.1 and 7.2.

Foot valve/reflex valve

The foot valve or reflux valve is an integral part of the pumping system. The foot valve is preferred while pumping water from open water , whereas reflux valve is used in

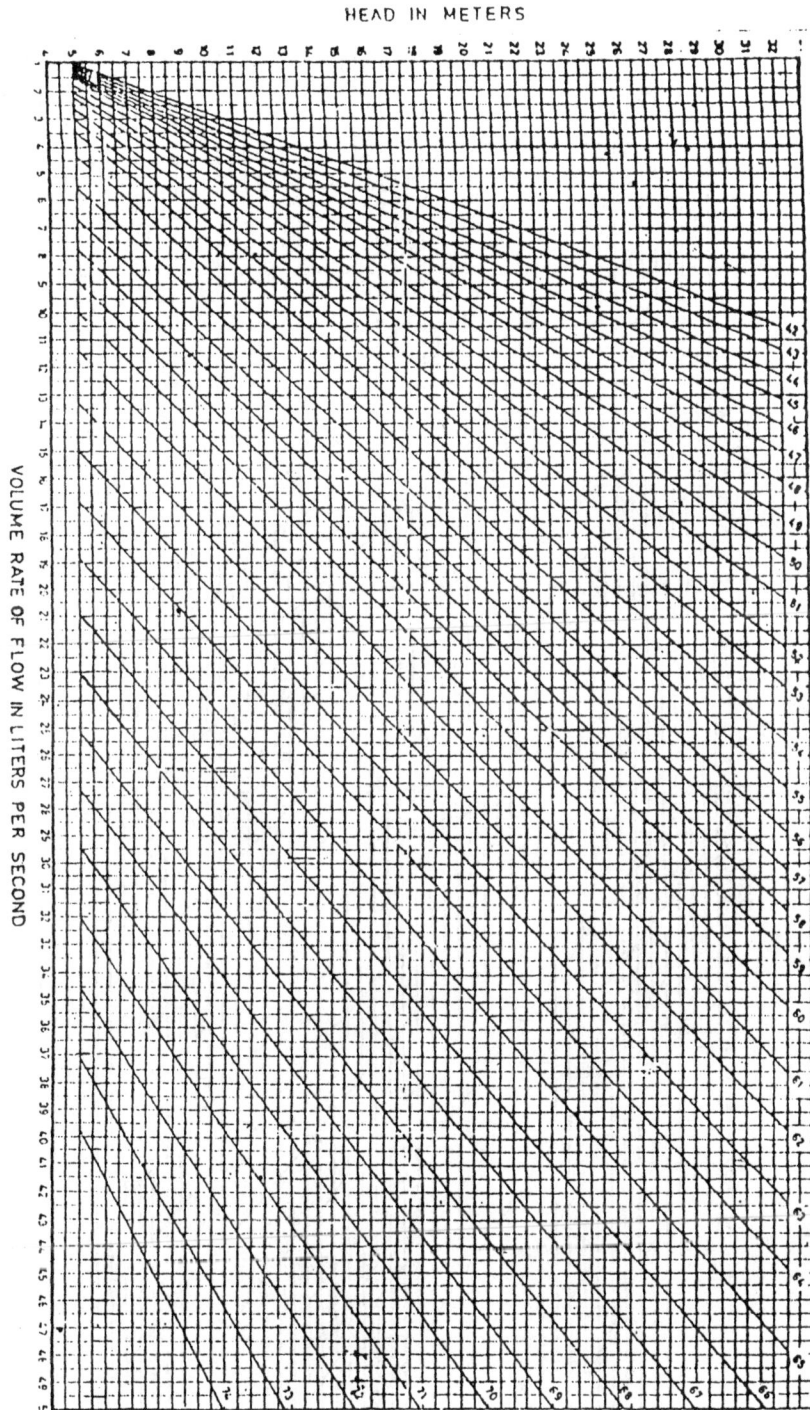

FIG. 7.1 EFFICIENCY IN PERCENT FOR HORIZONTAL CENTRIFUGAL PUMPS FOR AGRICULTURAL PURPOSES

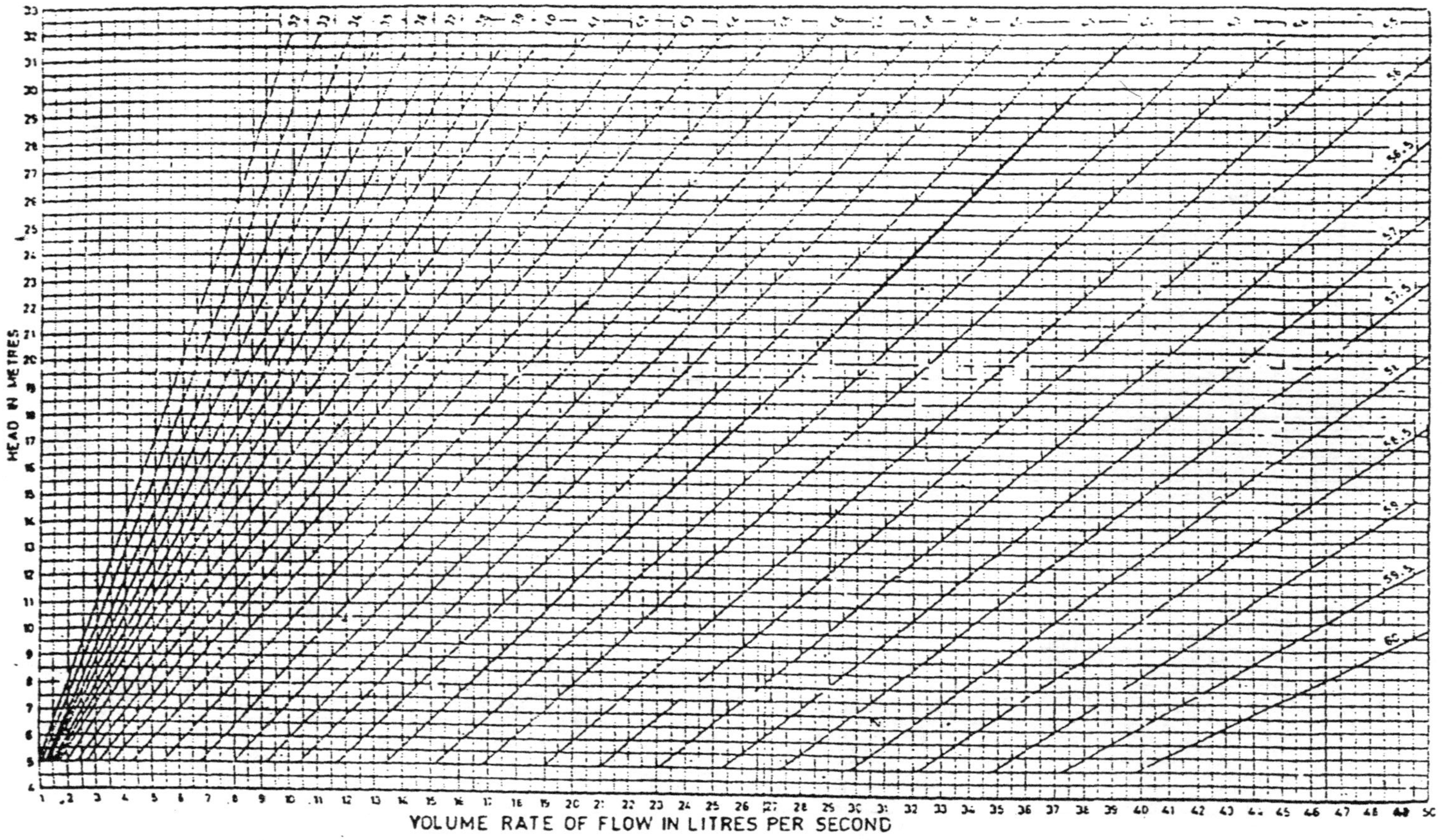

FIG.7.2 EFFICIENCY IN PERCENT FOR MONOSET PUMPS FOR AGRICULTURAL PURPOSES

case of tubewells. It plays an important role in the performance of the pumping system. Higher friction losses in the foot valve/reflux valve not only increase the total head of installation but it also increases the total suction lift, ultimately reducing the discharge and efficiency of pump. Friction losses in a foot valve/reflux valve are given as:

$$H = K \times V^2/2g$$

Where

H = Frictional losses
K = Coefficient of friction
V = Velocity of flow in suction pipe
g = acceleration due to gravity

To get the best performance from the pumping set, the valve of K should be as low as possible. Foot valves/reflux valve available in the market are having K valve ranging from 0.5 to 12. To give more discharge and consume less energy for delivering the same amount of water the valve of K should be less than one.

7.6 Data required for selection of centrifugal pump

For selecting the right size of the centrifugal pump for a particular set of conditions, the following information are required:

a) The source of water supply and fluctuations in the water level in summer and in the rainy season.
b) The vertical suction lift - the vertical distance from the water level to the centre of the pump.
c) The static discharge lift- the vertical distance from the centre of the pump to the delivery point.
d) The length of the delivery line in case the water is not discharged just outside the well.
e) The diameter of the tube-well.
f) The discharge capacity required.
g) The type of power-whether the electrical energy is available on the spot, or a diesel engine is to be used.

7.7 Quality criteria

Normally a complete pumping set is not manufactured by any manufacturer and there are separate Original Equipment Manufacturers for prime mover, pump, base frame, coupling and other accessories required to make a complete pumping unit. A manufacturer collects all such components, assembles the complete pumpset and provides it to the dealer.

a- Prime mover

I) Diesel engine:

Diesel engine is the most important and costly unit in the whole pumping system. A quality engine should be able to give the following results:

- i- Low specific fuel consumption
- ii- Low lub. oil consumption
- iii- Low exhaust smoke
- iv- High reliability and preferably should be IS mark

II) Electric motor:

It should preferably by IS mark. It is able to withstand a voltage variation from + 6 to -15% of rated voltage, which fairly covers the voltage variation of rated supply. The motor should have class E or higher class of insulation which are capable of withstanding temperature up to 115°C. The efficiency should not be less than 77% for 3 hp and 80% for 5 hp and above.

b- Centrifugal Pump

A good quality pump should have high reliability and high efficiency. Reliability is achieved by better material inputs, heat treatment, surface finish, proper design of bearing supports and appropriate tolerances etc. Pumps manufactured by many small units have faulty design of bearings and impellers, resulting in excessive friction and poor performance. Therefore, for selection of pump, the main consideration should be the overall efficiency and minimum consumption of power. A pump having minimum efficiency of 60% or above should be preferred.

c- Coupling of prime mover with pump

There should be minimum power loss transmission through coupling. The quality of rubber and canvass used in coupling should be of good quality. However, mono-pump set have the minimum losses and be preferred.

d- Suction condition and net positive suction head requirements

Water level variation between summer and monsoon seasons forces the farmer to use higher suction lift. The pump may start working with a suction lift of 3 metres and go on pumping upto 8 metres in most of the cases. This would result to drastic loss of efficiency in pumps with poor NPSH characteristics. NPSH requirements of the pump should be checked before the installation to avoid cavitation. Any leakage in the suction line would result to poor performance of the pump. The suction line should be airtight.

e- Suction and delivery hoses

Flexibility of hoses especially in suction hose is very important. It has been observed that cheaper quality hose gets cracked and results in leakage of water or clogging of inside surface due to unnatural union causing obstacle in the passage of water. A quality hose should have successfully tested for working temperatures, pressure and inside finish as per BIS standard.

f- Piping system

Generally farmers feel that if the pumps is discharging water with higher velocity they will get more water. This results invariably in use of smaller piping and foot valve, carrying a loss of nearly 25 to 30% due to friction. The size of pipe should be suitable for flow rate. The friction losses in pipe should not exceed 10% of total length of the pipe. Pipe material and inside surface should be as smooth as possible so that the pipe offers the least possible friction. There should be minimum possible number of bends in the piping system. Pipe fittings should match with the pipes. All the joints in the suction branch should be air tight.

Normally the farmer decides the purchase of a system based on the rating of the prime mover like 7.5 HP engine or a 5 Hp motor. The specifications of the pump, foot valve, reflex valve and pipings are neglected. He purchases an inferior prime mover. The pump, the valve and the piping normally do not match the requirement of his farm nor the prime mover. In addition these may be individually inefficient.

7.8 Selection of pump-set

It has been observed that the farmer's first choice is always for a quality pumpset but he is tempted to go in for sub-standard pumpsets only on price consideration. But it is desirable to recommend quality units to him even if it costs a little higher. Usually a pump is selected from the chart of total head v/s discharge rate supplied by the manufacturer. In this chart, the efficiency at different total heads is generally not indicated. So, either manufacturer should be asked to indicate efficiency at different total heads or to supply performance characteristics curves of the pump as shown in Fig.7.3. Draw the system head characteristics on the total head v/s discharge rate curve of the pump and find its intersection point with the total head curve for different pumps. Pump which gives highest efficiency and meeting the discharge requirements may be selected. A flat efficiency curve characteristic pump is always preferred. It should also be ensured that the pump is working in summer also, for both NPSH and total head requirements, as the water level goes down in summer. If a single pump is not sufficient to work in all seasons,a separate pump may be selected for summer season. If the yield of the well or hp consumption is a limiting factor, the period of operation may be adjusted suitably to irrigate the required area.

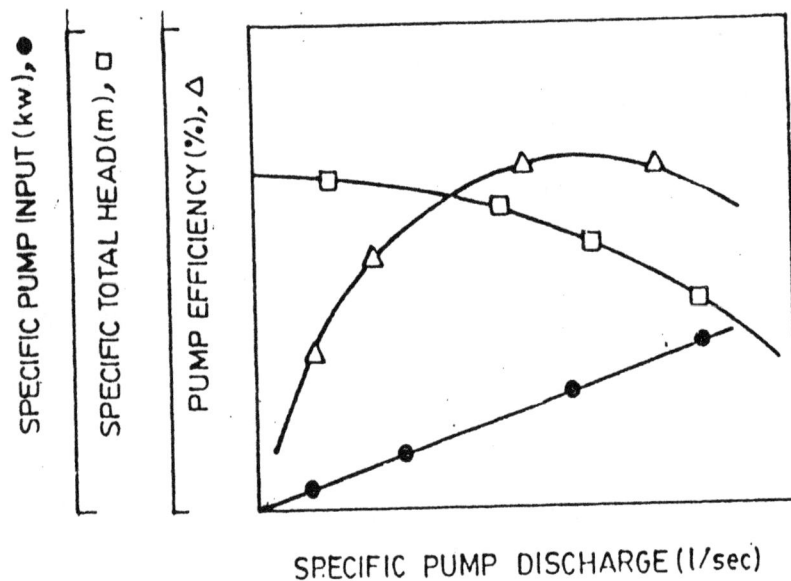

FIG.7.3 PUMP PERFORMANCE CURVE CHART

A Case Study on Selection of Pumpsets

Problem

A farmer has 7.5 hectares of land. He plans to grow sugarcane in 5.5 hectares and vegetables in balance 2.0 hectare. The source of water is a well and water table is 4.5 meters. The water is to be thrown at a height of 4.0 m. from the ground level. There is no draw-down problem in the well. Select the most suitable pump set.

Solution

i) Water requirement:

Total area to be irrigated	7.5 ha
Sugarcane area	5.5 ha
Vegetable area	2.0 ha
Irrigation interval for sugarcane	10 days
Depth of water required for sugarcane	10 cm
Irrigation interval for vegetable	10 days
Depth of water required for vegetable	2 cm

Area to be irrigated per day for sugarcane $\dfrac{5.5}{10} = 0.55$ ha/day

Water required for sugarcane $0.55 \times 10 = 5.5$ ha-cm/day

Area to be irrigated per day for vegetable $\dfrac{2}{10} = 0.2$ ha/day

Water requirement for vegetable $0.2 \times 2 = 0.4$ ha-cm/day

Total water requirement for sugarcane $5.5 + 0.4 = 5.9$ ha-cm/day
and vegetable

Corresponding discharge for 5.90 ha-cm/day   24 lit/sec
(Refer Appendix-VII)

ii) Suction and delivery line :

It is assumed that pump is of 100x100 mm size and flange is used to make it suitable
for 125x125 mm suction and delivery.(Refer to Appendix-VIII)

Static suction lift	4.5 m
Length of suction pipe	5.5 m (upto bend)
One medium type bend (dia-125 mm90° angle)	1.0 m
Length of suction pipe from bend to pump	2.6 m

It is assumed that dia of pipes and footvalve is 125 mm

Flange size (suction side)	100x125
Delivery head	4.0 m
Length of delivery pipe	5.5 m
One medium type bend is used	1.0 m
(dia. 125 mm and 90° angle)	
Flange size (delivery side)	100x125

Suction and delivery pipes are C.I. pipes

iii) Head loss in suction side:

Length of suction pipe	8.1 m
Head loss in foot valve (125 mm), equivalent to length of pipe in metre	10 m) Refer Appendix
Head loss in bend	3.66 m) IX.

Head loss in flange	1.0 m

Equivalent length of pipe for suction head loss in suction pipe, foot valve, bend and flange. 8.1+10+3.6+1=22.7 m

From the Table No.5 head loss at 24 lit/sec or 1440 lit/mt. .463 m/10 m

Actual head loss $\dfrac{0.463 \times 22.7}{10} = 1.01\,m$

Total suction head 4.5 + 1.01 =5.51 m

iv) Head loss in delivery side:

Length of delivery pipe	5.5 m
Head loss in bend	3.66 m (Refer Appendix-X (a)
Head loss in flange (100x125)	1.0 m
Total head loss equivalent to pipe length	5.5+3.66+1=10.16 m

From the Table No.5 head loss in meter for 24 lit/sec or 1440 lit/mt .463 m per 10 m pipe

Actual head loss $\dfrac{0.463 \times 10.16}{10} = 0.47\,m$

Total head delivery 4.0+0.47 =4.47 m

v) Total head = suction head + delivery head 5.51+4.47
9.98 m or 10 m (say)

A pump has to be selected for total head of 10 m and 24 lit/sec discharge keeping in view the best efficiency.

7.9 Guarantee and purpose of testing

The tests are intended to ascertain the performance of pump and to compare this with the Manufacturer's Guarantee. The following parameters are generally guaranteed:

a) Discharge rate
b) Total head
c) Pump input and pump efficiency
d) Net Positive Suction Head (NPSH)

It is necessary to specify the pump speed or the electrical supply frequency and voltage for the motor pump unit. Therefore, the guaranteed operational data shall form the basis of testing.

7.10 Testing procedure

Measurements shall be taken on not less than five different discharge values starting from full flow rate to nil discharge, and atleast one of them shall be measured at a head lower than the specified head. A suggested performa for observation and calculation is shown in Appendix-X (b) and X (c).

7.11 Testing parameters

1- Total head
2- Discharge rate
3- Specific pump input
4- Specific Pump output
5- Pump efficiency

The testing apparatus shall have adequate provisions for the testing of pump performance. Installation of test sample is shown in Fig.7.4 (a) and 7.4 (b).

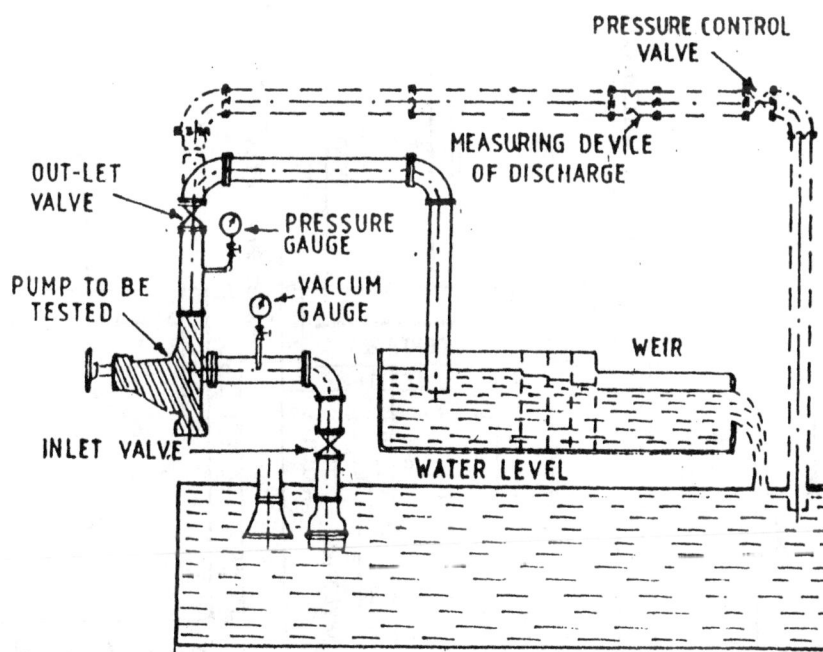

FIG.7.4(a) INSTALLATION OF TEST SAMPLE

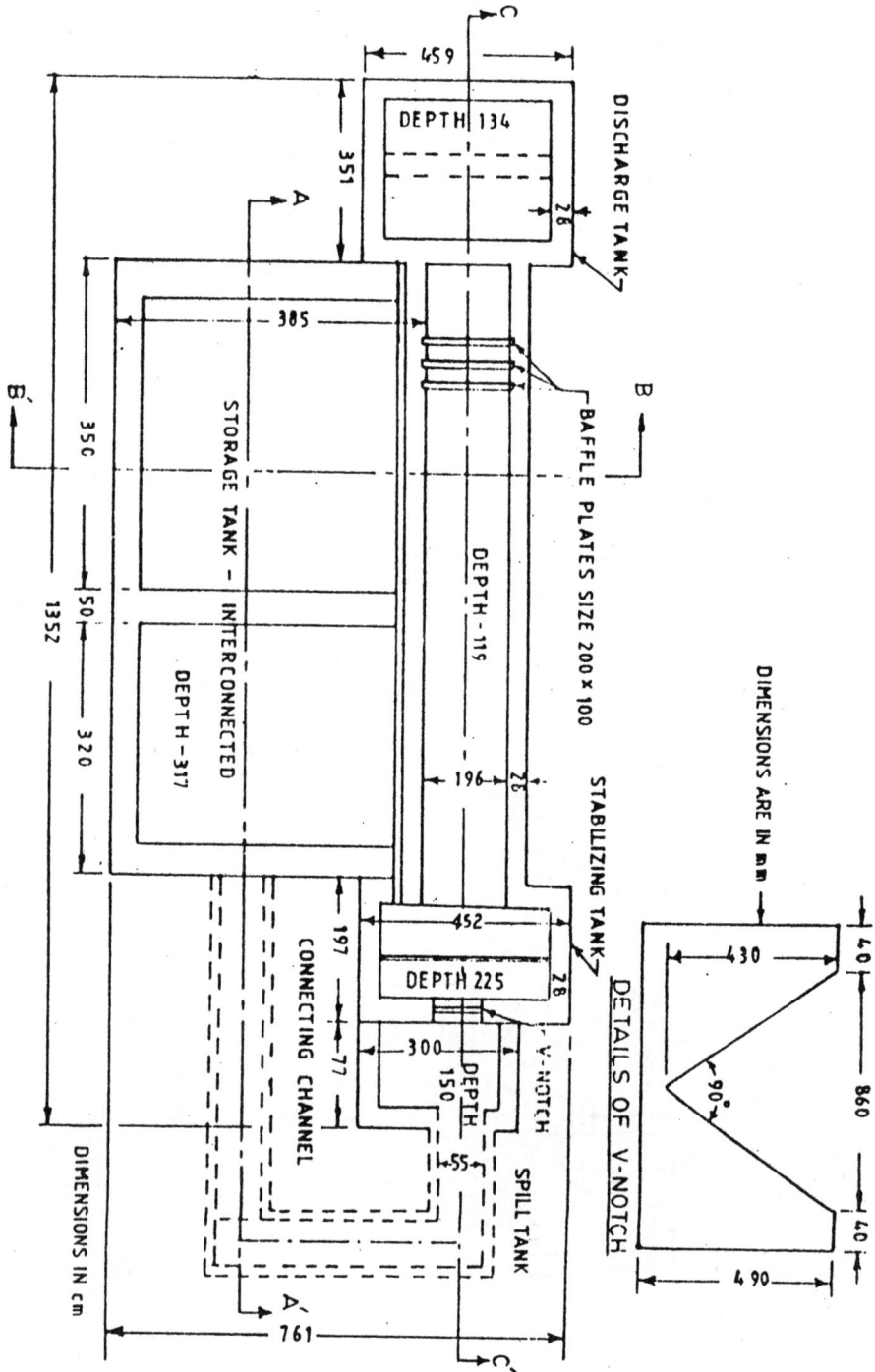

FIG.7.4(b) SUGGESTED CENTRIFUGAL PUMP DISCHARGE MEASUREMENT STRUCTURE

7.12 Measuring instruments

Total pump head

The measuring instruments used for the measurements of pressure shall be as follows:

1- Bourdon-tube gauge and vacuum gauge*
2- Liquid column gauge
3- U-tube mercury gauge
4- Pressure transducer*

*Before using, calibration of the gauge shall be made with a standard weight type pressure or standard liquid column type mercury pressure manometer.

Discharge rate

The measuring instruments used for the measurements of discharge rate shall be as follows:

1- Weirs : i) Right-angle triangular weirs
        ii) Rectangular weirs
        iii) Full-width weirs
2- Float type area flow meter*
3- Electromagnetic flow meter*
4- Turbine type flow meter*

* The flow meter shall be calibrated before use, and the calibration shall be performed in accordance with weight method, by using a sensor or volume method by using a means of container.

Shaft power requirement

The shaft power requirement shall be obtained by measurement of the input of the driving motor of known characteristics through an accurate testing or by using dynamometer.

Measurement of speed of rotation

The speed of rotation shall be measured directly as far as possible with the help of tachometer.

Accuracy of measurement

The measuring instruments should not exceed the systematic error limit specified below as well as the overall error limit.

Parameters	Permissible error of measuring instruments(%)	Permissible overall error(%)
i) Rate of flow, total head	±2.5	±3.5
ii) Pump power input		
Electric power input	±2.0	±3.5
iii) Speed of rotation	±1.4	±2.0
iv) Motor efficiency	±2.0	-
v) Pump efficiency		±5.0

7.13 Calculation of performance parameters

a- Discharge with 'V' notch

$$\text{Discharge lit/sec} = \left(\frac{\text{Head over notch in cm}}{20600}\right)^{2.48}$$

b- Velocity head (suction or delivery side)

a) Pipe area $\quad \frac{\pi}{4} d^2$

b) Velocity (v) $= \dfrac{\text{Discharge}}{\text{Area}}$

c) Velocity head $= \dfrac{v^2}{2g}$ (Where g is acceleration due to gravity)

c- Total head = Suction head + delivery head + manometer distance
+ velocity head (delivery side) - velocity head (suction side)

d- Speed factor(N) $= \dfrac{\text{Actual measured speed}}{\text{Specified declared speed}}$

e- Specific discharge $= \dfrac{\text{Actual measured discharge}}{N}$

f- Specific head $= \dfrac{\text{Actual measured head}}{N^2}$

$$\text{g- Specific motor input} = \frac{\text{Actual measured horse power}}{N^3}$$

h- Specific pump input = (Sp. motor input x motor efficiency)-transmission losses
(Transmission loss should be taken as 6% in flat belt & 3% in 'V' belt)

$$\text{i- Specific pump output or specific water hp} = \frac{\text{Specific discharge (Lit/sec.) x sp. head (m)}}{76}$$

$$\text{j- Pump efficiency (h)} = \frac{\text{Specific pump output}}{\text{Specific pump input}} \times 100$$

7.14 Case study for analysis of testing data

A pump of size 100x100 mm was tested with motor coupled by flat belt. Different readings were taken by throttling delivery valve. Discharge was measured with 'V' notch. Specified speed of pump is 1450 rpm. Guarantee point declared by manufacturer are Head= 21m, Discharge= 28 lit/sec. Efficiency 64%, Observed data for one set is as follows:

Actual speed of pump	= 1440 rpm
Suction head (vacuum gauge)	= 4.5 m
Delivery head (Pressure gauge)	= 14.0 m
Manometer distance i.e. distance between two gauges	= 0.725 m
Hook gauge initial reading at 'V' notch	= 354.0 cm
Hook gauge final reading at 'V' notch	= 128.5 cm
Voltage during testing	= 400 volt
Current during testing	= 26.2 amperes
Motor input reading	= 11.3 KW
Motor efficiency after calibration	= 85.2%

Calculate the performance of pump
Solution:

Head over 'V' notch = 354-128.5 = 225.5 cm

$$\text{Discharge} = \left(\frac{225.5}{20600}\right)^{2.48} = 33.26 \text{ lit/sec}$$

$$= 0.03326 \text{ m}^3/\text{sec}$$

$$\text{Suction pipe (cross sec. area)} = \frac{\pi D^2}{4} = \frac{\pi \times 100^2}{4 \times 1000} = 0.00785 \text{ m}^2$$

$$\text{Velocity (V)} = \frac{\text{Discharge}}{\text{Area}} = \frac{0.03326}{0.00785} = 4.23 \text{ m/sec}$$

$$\text{Velocity head (suction)} = \frac{V^2}{2g} = \frac{(4.23)^2}{2 \times 9.81} = 0.914$$

Velocity head (delivery) = 0.914 (Since suction & Delivery pipe dia are same)

Total head = Suction head + Delivery head + Manometer distance+Velocity head (delivery) - Velocity head (suction)

$$= 4.5 + 4.0 + 0.725 + 0.914 - 0.914$$
$$= 19.22 \text{ m}$$

$$\text{Speed factor (N)} = \frac{\text{Actual speed}}{\text{Specified speed}} = \frac{1440}{1450} = 0.993$$

$$\text{Specific discharge} = \frac{33.26}{0.993} = 33.49$$

$$\text{Specific head} = \frac{19.22}{(0.993)^2} = 19.49$$

$$\text{Specific pump input} = \frac{11.3}{(0.99?)^3} \times 0.852 = 9.83 \text{ KW}$$

$$\text{Specific pump input (after deducting 6\% belt losses)} = \frac{9.83 \times (100-6)}{100} = 9.24$$

$$\text{Specific pump output or Sp. water hp} = \frac{33.49 \times 19.49}{76} = 8.59 \text{ hp}$$

$$\text{Specific pump output (KW)} = \frac{8.59 \times 735.5}{1000} = 6.31 \text{ KW}$$

$$\text{Pump efficiency} = \frac{6.31 \times 100}{9.24} = 68.29\%$$

Similarly other sets of reading are calculated and characteristics curves are drawn as shown in Fig. 7.3.

7.15 Verification of guaranteed duty points

A) Head and Discharge

The guarantee duty point declared by the manufacturer i.e. discharge (QG) and Head (HG) are plotted as straight line on the H-Q curve of the pump undergone testing as shown in Fig.7.5. The point is extended vertically and horizontally so that the lines intersect on the drawn curve. Then measure the distance DH and DQ from the test curve. Tolerances ± XQ and ± XH respectively shall be applied to the guaranteed duty point QG and HG. In general the following values may be used:

$$X Q = 0.07$$
$$XH = 0.04$$

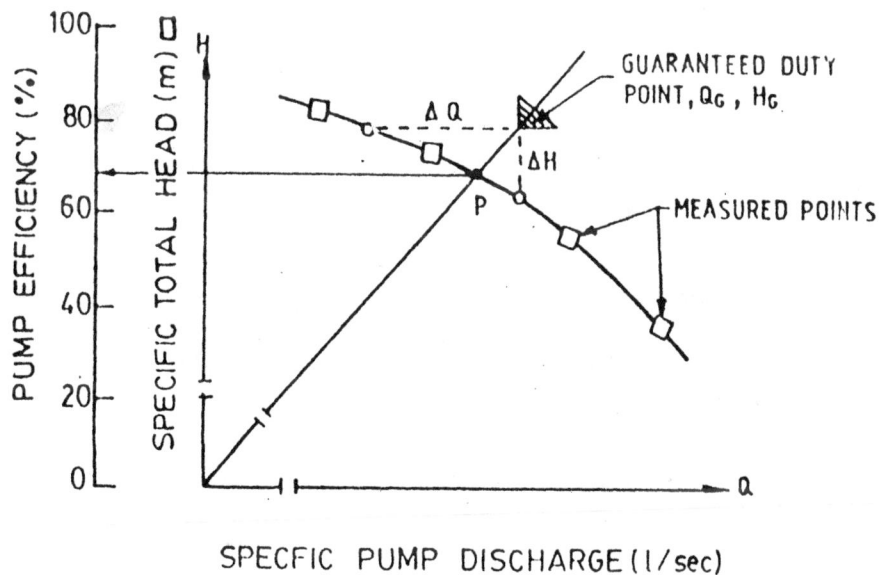

FIG.7·5 VERIFICATION OF GUARANTEE POINTS

108

The following test formula is applied for verification of a guarantee:

$$\left(\frac{HG \cdot XH}{\Delta H}\right)^2 + \left(\frac{QG \cdot XQ}{\Delta Q}\right)^2 \geq 1$$

If the calculated value is greater than or equal to 1, the guarantee condition will be considered as qualified and if the value is less than 1, the guarantee condition not fulfilled.

B) Efficiency :

The efficiency shall be derived from the drawn Q-H curve where it is intersected by the straight line passing through the declared duty point $Q_G H_G$ at point P as shown in Fig.7.3. The efficiency at the point of intersection shall be atleast 95% of the specified value as the maximum permissible overall error being ± 5%. But for combined motor pump units the maximum permissible overall error will be ± 4.5%.

Example for verification of guaranteed duty point:

Declared duty point:

    Head (HG) = 21 m
    Discharge (QG) = 28 lit/sec
    Efficiency = 64%

Calculation :

    X H = 0.04) Standard values
    X Q = 0.07)

    DH = 0.6
    DQ = 0.9 Measured from test graph

$$\left(\frac{H G \times 0.04}{\Delta H}\right)^2 + \left(\frac{Q G \times 0.07}{\Delta Q}\right)^2 \geq 1$$

$$\left(\frac{21 \times 0.04}{0.6}\right)^2 + \left(\frac{28 \times 0.07}{0.9}\right)^2 = 6.7$$

                       Which is greater than 1

Hence guarantee point confirms
Declared efficiency = 64%

$$\text{After permissible error} = \frac{64 \times 95}{100}$$

$$= 61.8\%$$

Derived from Q-H curve = 66%
Which is greater than declared. Hence confirms.

Chapter 8

TESTING AND EVALUATION OF PLANT PROTECTION EQUIPMENT

8.1 Introduction

The importance of plant protection measure for stepping up agricultural production has been widely accepted all over the world. Among the various methods of pests control. chemical method is most effective. Chemical pesticides have played and will continue to play a major role in the rapid advancement of agriculture. In addition to improvement in crop growth, quality and yield through proper application of specified insecticide and pesticide. the use of chemical herbicides has reduced labour intensive requirements for weed control. Effective control of injurious organisms is impossible without efficient and appropriate plant protection equipments, through which right dose of chemical is to be applied on plant surfaces in the form of sprays, dust. mist etc. Equipments are required generally for the following applications:

i. Application of insecticide to control insects on plants
ii. Application of fungicides to control plant diseases
iii. Application of herbicides to destroy weeds.
iv. Application of hormone for growth regulation in order to increase fruit set or prevent early dropping of fruit
v. Application of plant nutrients directly to the plant foliage
vi. Application of biological materials such as viruses and bacteria in sprays to control insects/pests.

8.2 Type of pesticide application equipment

1. Manually operated Hand sprayer atomizer
2. Hand compression sprayer
3. Knapsack sprayer
4. Foot Sprayer
5. Rocker Sprayers
6. Dusters
7. Power sprayers
    i) Hydraulic power sprayers
    ii) Pneumatic power sprayer-cum-duster
8. Tractor mounted Power-take-off operated sprayer

Manually operated hand sprayer or atomizer

The manually operated hand sprayer or atomizer is a small, light and compact unit. The capacity of the container is about 500 ml. It is meant for small spraying jobs in and around the house,e.g. spraying small flower beds and vegetable plots in kitchen garden.

Hand compression sprayer

The typical hand compression sprayer comprises of a cylindrical tank for holding the spray fluid with a handle, filler hole, spray lance, nozzle and cut-off device. The capacity of the tank varies from 10-20 litres.

Knapsack sprayer

The type commonly available in India is the lever-operated plunger or diaphragm type. It has a flat bean-shaped tank. The body of the sprayer is so shaped so as to conveniently fit it on the back of the operator. The capacity of the container is from 10-15 litres. It is generally made of brass or plastic. In some cases it is provided with a built-in double barrel pump of piston type, or of diaphragm type, with a lever for operating it.

Foot sprayer

The foot sprayer consists of a plunger assembly, a stand, a suction hose, a delivery hose, an extension rod with a spray nozzle, etc. One end of the suction hose is fitted with a strainer and the other with a flexible coupling. Similarly the delivery hose has one end fitted with a cut-off valve and the other with a flexible coupling. It is operated by foot. The pump is fixed on an iron stand, and a pedal attached to the plunger rod, operates the sprayer by its upward and downward movements. It does not have a built-in tank. It is used for crop and fruit trees up to 4 meters in height. It may or may not be mounted on a trolley.

Rocker sprayer

The rocker sprayer consists of a pump assembly, a platform, an operating lever, a pressure chamber, a suction hose with a strainer, a delivery hose , an extension rod with a spray nozzle, etc. The rocking movement of the handle operates the pump, which results in building up pressure in the pressure chamber. There is no built-in tank. Therefore, a separate spray tank is necessary. A high pressure can be built up in the tank. It can, therefore, be used for spraying tall field crops and trees upto 5 meter in height.

Duster

Appliances that are used for distributing dust formulation are called dusters. All machines used for applying dust consist essentially of a hopper (dust chamber) which usually has an agitator in it, an adjustable orifice or other metering mechanism and delivery tube. The rotary fan supplies the air stream. Generally two types of dusters are available namely the plunger type and the crank or rotary type.

Power sprayers

i) Hydraulic power sprayer

These are based upon utilizing hydraulic energy for atomizing and spraying the liquid. They are fitted with any of the pumps, namely, piston type, plunger type, roller vane type, diaphragm type, gear type or centrifugal type.

ii) Pneumatic power sprayer-cum-duster

These are based upon pneumatic energy. The pneumatic air pressure is thus utilized to agitate the spray or dust. Pneumatic sprayers can be of knapsack type which is mounted on the back of operator & barrow type which is mounted on wheel barrow. While using this sprayer as duster, one has to close the spray delivery tube and open the dust delivery tube, set the air supply for dusting and then start engine for dusting.

Tractor mounted pto operated sprayer

It can be used for plant protection, weedicides spraying and liquid fertilizing. It consists of a centrifugal type tulu pump, a tank and booms of flexible hose pipes on which nozzles are fixed. This boom is tied up on rigid beam by clamps. Spacing of the nozzles on the boom can be adjusted to suit the row crop spacing by fixing clamps at desired place. A pressure gauge with a pressure regulator valve is provided to control the spray pressure. The pump is driven from tractor pto. Machine can cover a 2 to 3 ha/hr at an operational speed of approx. 5 km/h depending upon the field size and field conditions.

8.3 Definition of Important Terms

The following definitions shall apply in the test code and procedure :

a- Continuous Sprayer- A sprayer in which the pump has to be operated continuously while liquid is discharged.

b- Piston type - A continuous sprayer, the pump of which is piston or plunger type.

c- Continuous sprayer- diaphragm type- A continuous sprayer, the pump of which is diaphragm type.

d- Compression sprayer- A sprayer, the liquid tank of which is a pressure vessel and in which the discharge is carried out with the air pressure created in advance by built-in pump or from outside.

e- Compression sprayer- non-pressure retaining type : A compression sprayer in which the working pressure does not remain constant but decreases gradually during discharge.

f- Pressure chamber: It is a component to even out the fluctuations of the liquid pressure and maintains uniform flow of liquid.

g- Strokes - The maximum travel of the piston or plunger rod in one direction when the handle moves from its lowest possible position to its highest possible position.

h- Tank capacity- The volume of the tank when liquid is filled to its neck level.

i- Volumetric efficiency : It is the ratio of actual volume of the spray fluid discharged to the piston or plunger displacement in one stroke. It is expressed in percentage.

j- Knapsack sprayer : A sprayer which can be mounted on the back of an operator for spraying.

k- Shoulder sprayer : A sprayer which can be suspended from the shoulder of operator for spraying.

8.4 Type of tests for sprayers

1- General Tests
    a- Visual examination
    b- Checking of dimensions
    c- Checking of material of construction

2- Performance tests
    a- Discharge rate
    b- Volumetric efficiency
    c- Pressure development
    d- Pressure retention
    e- Liquid remains
    f- Leakage
    g- Spray throw
    h- Spray coverage

3- Component tests
    a- Pressure chamber, pump cylinder and pressure tank
    b- Hose and hose connection
    c- Strap and its assembly
    d- Gasket test

114

e- Spring test
f- Operating lever, handle and piston rod for Knapsack srayer
g- Handle, piston rod, foot rest and stirrup for stirrup sprayer
h- Frame, piston rod, pedal lever or handle lever and extension for foot and rocker sprayer
i- Frame, operating lever, connecting rod and handle for charge pump
j- Valve assembly test
k- Tank impact test
l- Fatigue test
m- Test for spray lance

4- Performance test for cut-off device

5- Performance test for nozzle

6- Endurance test

8.5 Testing method for sprayers

1- General tests

a- Visual examination:
The equipment shall be visually examined with respect to the functional require ments given by the manufacturer. Conformity or otherwise of the requirements shall be reported.

b- Checking of dimensionsal measurements will be taken, compared as per relevant Indian standards and reported accordingly.

c- Checking of material of construction:
The material of constructions of various parts of the sprayer shall be checked as per codes and reported.

2- Performance tests

a- Discharge rate:
This test is applicable to continuous knapsack sprayers, foot sprayers, rocker sprayers and stirrup type sprayers. This test is conducted on a standard test rig as shown in Fig.8.1 having crank mechanism to operate sprayer at the speed of 16 ± 1 cycle per minute. The discharge is measured at a relevant pressure norms.

b- Volumetric efficiency:
This test is applicable to continuous knapsack sprayer, foot sprayer, rocker sprayer and stirrup sprayer. The discharge of water in 10 successive cycles is collected and measured at the same speed of 16 ±1 cycle per minute on a standard test rig Fig. 8.1.

115

1. MOTOR 5 H P   3. CRANK   5. VALVE
2. GEAR BOX   4. PRESSURE GAUGE   6. PLATFORM
7.   - ANGLE-IRON FRAME

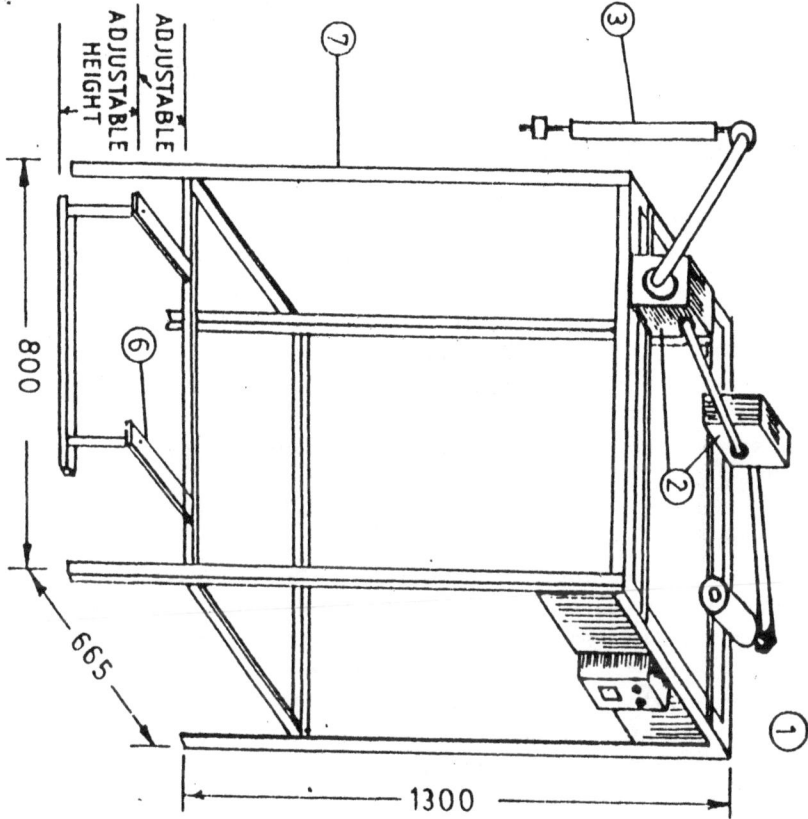

ALL DIMENSIONS ARE IN mm

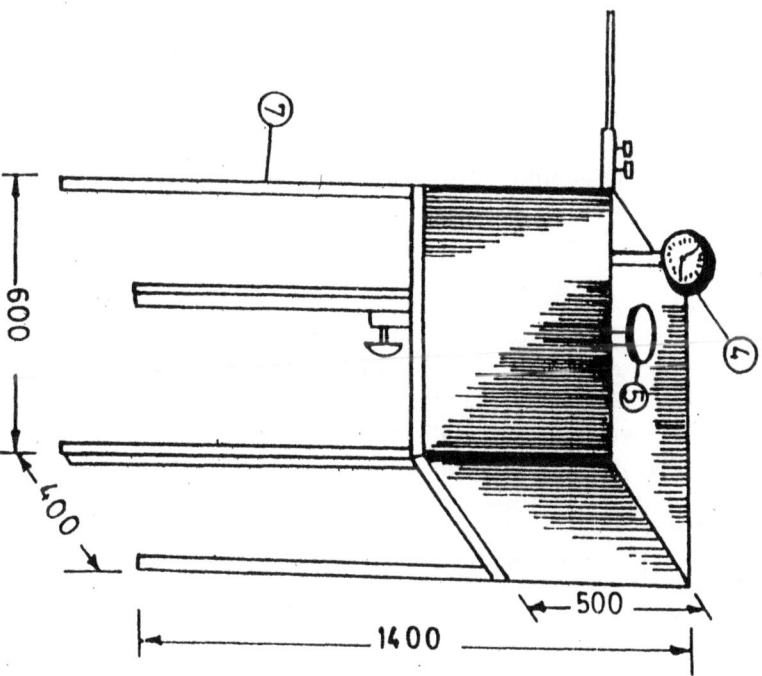

FIG. 8.1 PUMP PERFORMANCE AND ENDURANCE TEST RIG (ISO-VIEW)

The test is replicated four times and average value of discharge is taken. Accordingly, the volume of water discharge in one cycle is calculated. This is called actual discharge of sprayer. In order to calculate the theoretical discharge, the piston displacement is calculated by measuring the inner diameter of pump cylinder and the actual length of one stroke. Dividing the actual discharge by theoretical discharge we get volumetric efficiency copared with the norms and reported.

c- Pressure development:
This test is applicable to compression knapsack sprayer and is designed to check the pressure development in tank. The tank of sprayer is filled with clean water upto 2/3rd of its total capacity. The discharge outlet is closed and pressure gauge is fitted on the tank to read the pressure. The handle of sprayer is operated continuously 100 times at a constant speed on the rig as shown in Fig.8.2. The pressure so developed in the tank is recorded from the pressure gauge. This test should be repeated four times and average value is reported. If the minimum average value is 400 KPa then it conforms to Standard.

d- Pressure retention.
This test is applicable to compression knapsack sprayer (pressure retaining type) in order to check the pressure retention in the tank. The tank of sprayer is initially charged with an air pressure of 275 KPa. The desired quantity of liquid is then pumped inside the tank at a pressure of 833 KPa by charge pump. The liquid is discharged without using the nozzle upto its maximum possible extent. This test is repeated 100 times and at the end of each test. reading of air pressure is taken and average is reported. For the purpose of routine and acceptance check the sprayer is subjected to a minimum of 5 repetitions instead of 100 times.

e- Test for liquid remains:
This test is conducted for compression knapsack sprayer (pressure retaining type). Initially the sprayer is charged with water by applying two to three strokes of charge pump. Then conduct the operation as given in test for pressure retaining. When it is not possible to take out liquid. the air pressure shall be released from pressure release device. Then the amount of liquid left over, if any. in the tank shall be measured and reported.

f- Test for leakage:
This test is applicable to atomizer sprayer. Initially the piston assembly and stopper is removed and the outlet of sprayer as well as reservoir filler hole is sealed. A hose is fitted to the opening of pump cylinder and is pneumatically pressurized to an internal pressure of 34 KPa. The pressure is retained for a period of 2 minutes and the sprayer is immersed in water. During this test, if the sprayer does not show any sign of leakage. the sprayer is deemed to have passed this test.

q- Test for spray throw:
This test is applicable for atomizer sprayer, in order to check the spray throw. The

1. MAIN SWITCH   2. STARTER
3. GEAR BOX   4. CRANKS
5. LOCK   6. SPRAYER PLATFORM
7.   ANGLE-IRON FRAME.

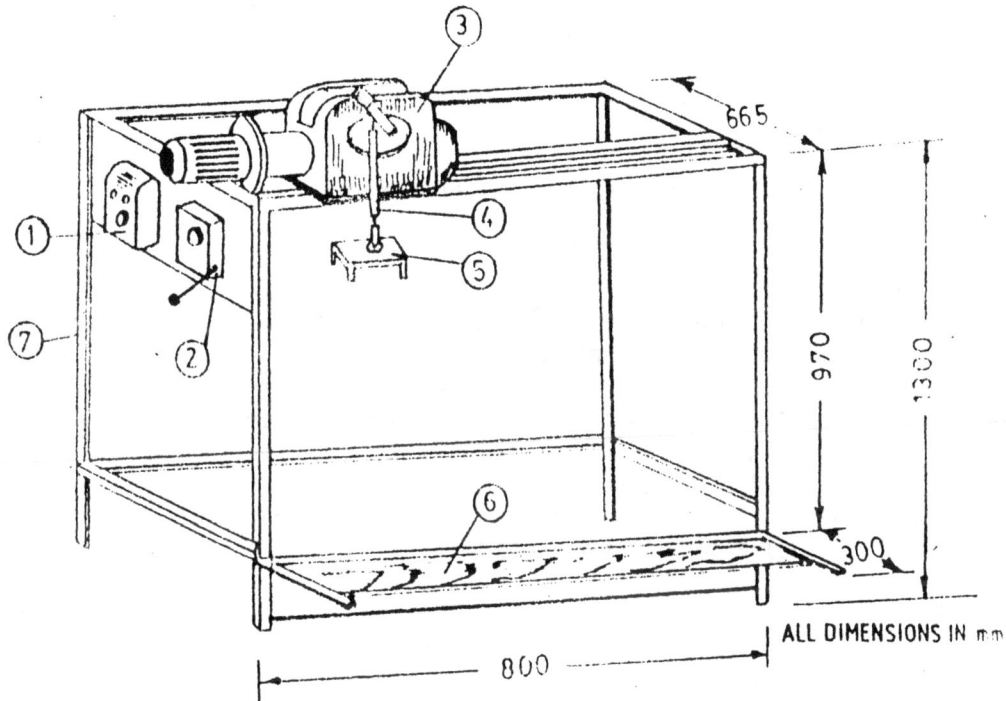

ALL DIMENSIONS IN mm

FIG. 8.2 TEST RIG FOR COMPRESSION SPRAYER ENDURANCE TEST

sprayer is firmly secured. An area phased with blotting paper or any other suitable paper and one metre in diameter on vertical sides at a distance of 1.5 metre from the tip of the nozzle of the sprayer is placed. The reservoir of sprayer shall be filled with kerosene oil to 2/3rd of its capacity. A small quantity of suitable colouring material such as waxoline which should not affect the viscosity of the kerosene oil materially should be added. The discharge of sprayer should be 12 ± 2 ml of oil per minute at an approximate speed of 60 cycles per minute. The sprayer shall be allowed to continue for a period of about 2 minutes. The sprayer shall be deemed to have passed this test if the spray reaches the paper.

h- Test for spray coverage:
This test is also applicable for atomizer sprayer in order to check the area of spray impression. The procedure for test is followed as given in test for spray throw except

that the distance of paper from the tip of nozzle should be kept as 545 mm. The spray shall be allowed to continue for a period of about 15 seconds. The area of spray impression on the paper shall be measured within a period of 5 minutes after completion of the spray. The sprayer shall be deemed to have passed this test if the area of spray impression is not less than 50 cm^2.

3- Component tests

a- *Test for leakage and deformation of pressure chamber, pump cylinder and pressure tank*

Under this clause pneumatic test and hydraulic tests are conducted. The procedure is given below :

i. Pneumatic test

A hose shall be fitted to the opening of the pressure chamber or pump cylinder or tank. In case there are more openings, all will be sealed except to which the hose is fitted. The pressure chamber or pump cylinder or tank shall be then pneumatically pressurized to a minimum of 1.5 times the normal working pressure of the sprayer. The pressure shall be retained for a period of one minute. The component shall be disconnected, immersed in water and examined for any leakage and deformation. The pressure chamber or pump cylinder or tank shall deemed to have passed this test if no leakage or deformation is found during this test.

ii. Hydraulic test

The sprayer tank shall be filled with water upto its capacity. The hose connections are to be made as per pneumatic test. Then the pressure chamber or pump cylinder or tank shall be pressurized to a static hydraulic pressure of a minimum two and a half times the normal working pressure of the sprayer. The pressure shall be retained for a period of one minute. The pressure chamber or pump cylinder or tank shall be deemed to have passed this test if no leakage or deformation is found during the test.

b- *Test for hose and hose connection*

The inlet of the hose pipe shall be connected to a hydraulic pump through hose connection. The other end of hose pipe shall be connected to the cut-off device. The outlet of the cut-off device shall be closed in such a way that no discharge is allowed. Aluminum hydrostatic pressure of 1.5 MPa, using water as a liquid shall be developed in the hose assembly and retained for a period of one minute. The hose and hose connection shall conform this test if no leakage or crack is observed during the test.

c- *Test for strap and its assembly*

This test is applicable to knapsack sprayer in order to check the strength of strap and its assembly on the test rig as shown in Fig. 8.3. The tank is filled with clean water

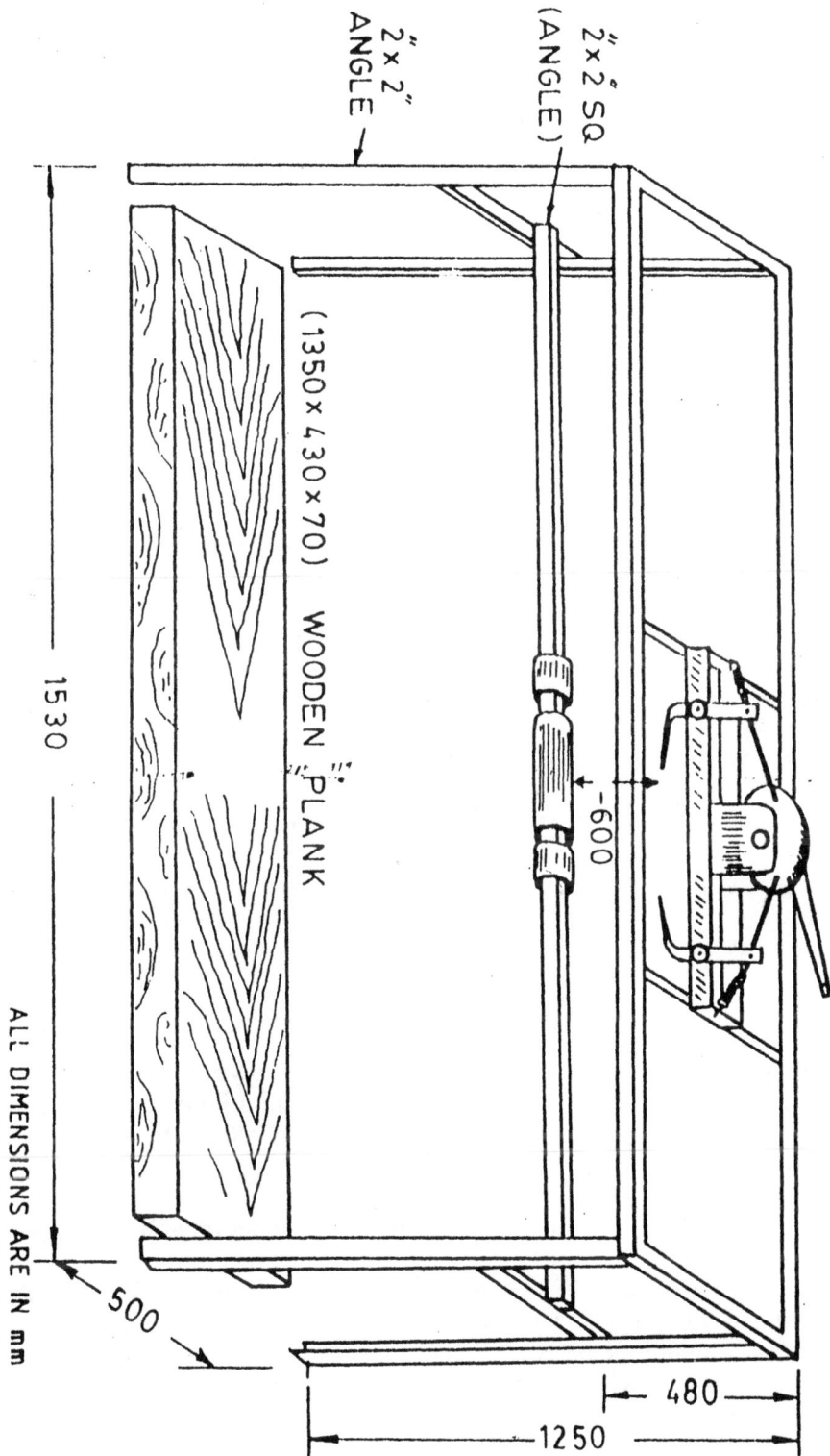

FIG.8.3 STRAP DROP AND TANK IMPACT TESTING RIG

120

to its specified capacity. The sprayer is hung from a solid support through its strap, simulating the conditions on the shoulder of the operator. Raise the tank vertically to a height of 300 mm and allow to drop freely. Repeat the operation 24 times. The assembly conforms test if no part of strap, bracket or clamp etc. is broken.

d- Gasket test

A new set of gasket of the sprayer shall be immersed in a mixture consists of 60% kerosene, 5% benzene, 20% toluence and 15% Xylene for a period of 72 hours at a temperature of 27 to 33 degree C. Then gasket are dried in air at same temperature range for a period of 24 hours. Then these gaskets be fitted in their original positions on sprayer. The sprayer complete with its discharge line shall be operated at its normal working speed and conditions for 8 hours. The gasket shall conform this test if no leakage is observed.

e- Spring test

This test is applicable for the springs provided in foot sprayer. The free length of spring is measured. The spring shall be attached to test rig in vertical position. It shall be compressed upto touching position of coil at a rate of 20 stroke per minutes for a period of one hour. The free length of spring shall be measured after test. The spring shall confirm this test if the difference in free length of the spring does not exceed 5%.

f- Test for operating lever, handle and piston rod for knapsack sprayer

This test is applicable to knapsack sprayer piston type in order to check their strength. The discharge outlet of sprayer is closed and handle is operated to develop the pressure in the sprayer to the minimum two and half times the normal working pressure. When the handle, operating lever and piston rod are operated at this pressure, there should not be any distortion or crack in order to confirm this test.

g. Test for handle, piston rod, foot rest and stirrup

This test is applicable in case of stirrup type sprayer in order to check the strength of components. The procedure to carry out is same as given in test for operating lever, handle and piston rod. This test deemed to have confirmed if no breakage or deformation is occurred.

h. Test for frame, piston rod, pedal lever or handle lever and extension

This test is applicable in case of foot and rocker sprayers. The procedure for testing is same as for operating lever, handle and piston rod test. Frame, piston rod and pedal lever in case of foot sprayer whereas frame, piston rod, handle lever and extension in case of rocker sprayer shall not break, deform or crack when the force is exerted on them in order to pass this test.

i. *Test for frame, operating lever, connecting rod and handle*

This test is conducted for charge pump. The pump outlet is closed so that no discharge is allowed from the pump. The handle is operated till a minimum pressure of 2 MPa is developed in the pump. These parts shall not break, deform or crack when the force is exerted at this pressure by moving the handle in order to pass this test.

j. *Valve assembly test*

This test is applicable for compression knapsack sprayer. The piston rod and piston assembly is taken out. The discharge outlet is connected to a hydraulic pump through hose connection. The outer openings except discharge are sealed. The tank shall be pressurized to a static hydraulic pressure of minimum two and a half times the normal working pressure of the sprayer. This pressure is retained for a period of five minutes. The valve assembly is confirmed if no drop in pressure is observed.

k. *Tank impact test*

This test is applicable in case of compression knapsack sprayer and is conducted on test rig as shown in Fig.8.3. The tank is filled with clean water upto 2/3rd of its capacity and pressurized pneumatically to the normal working pressure of the sprayer. Take out the discharge line and plug the discharge outlets. The tank shall be dropped for 25 times from a height of 600mm in following positions :

i- 7 times with its long axis horizontal
ii- 6 times with the long axis vertical
iii- 6 times with the long axis inclined at 75" to the horizontal
iv- 6 times as in (iii) with the position of impact diametrically opposite

The platform on which the tank is dropped shall consist of a plane solid teak wood or similar hard wood of 60 mm thick, placed on a hard level surface. The sprayer is deemed to have passed this test if tank shall not burst and bottom of tank or any other part of sprayer shall not extend below the bottom of skirt.

l- *Fatigue test*

This test is conducted for compression knapsack sprayer on test rig as shown in Fig. 8.4. The pump assembly and discharge line are taken out from the sprayer. An opening of the tank is connected to the manifold of test rig through a hose and shut-off cock. The tank is completely filled with clean water. The manifold is also filled with clean water through the filler hole. Air pressure with the help of compressor shall be developed equal to the normal working pressure of the sprayer within the range of ± 10 through a three way valve. A timer switch connected in the circuit is used to open and close the three way valve within a range of 3 to 5 times per minute. The tank is subjected to 1200 such pressure cycles. A counter connected in the system will

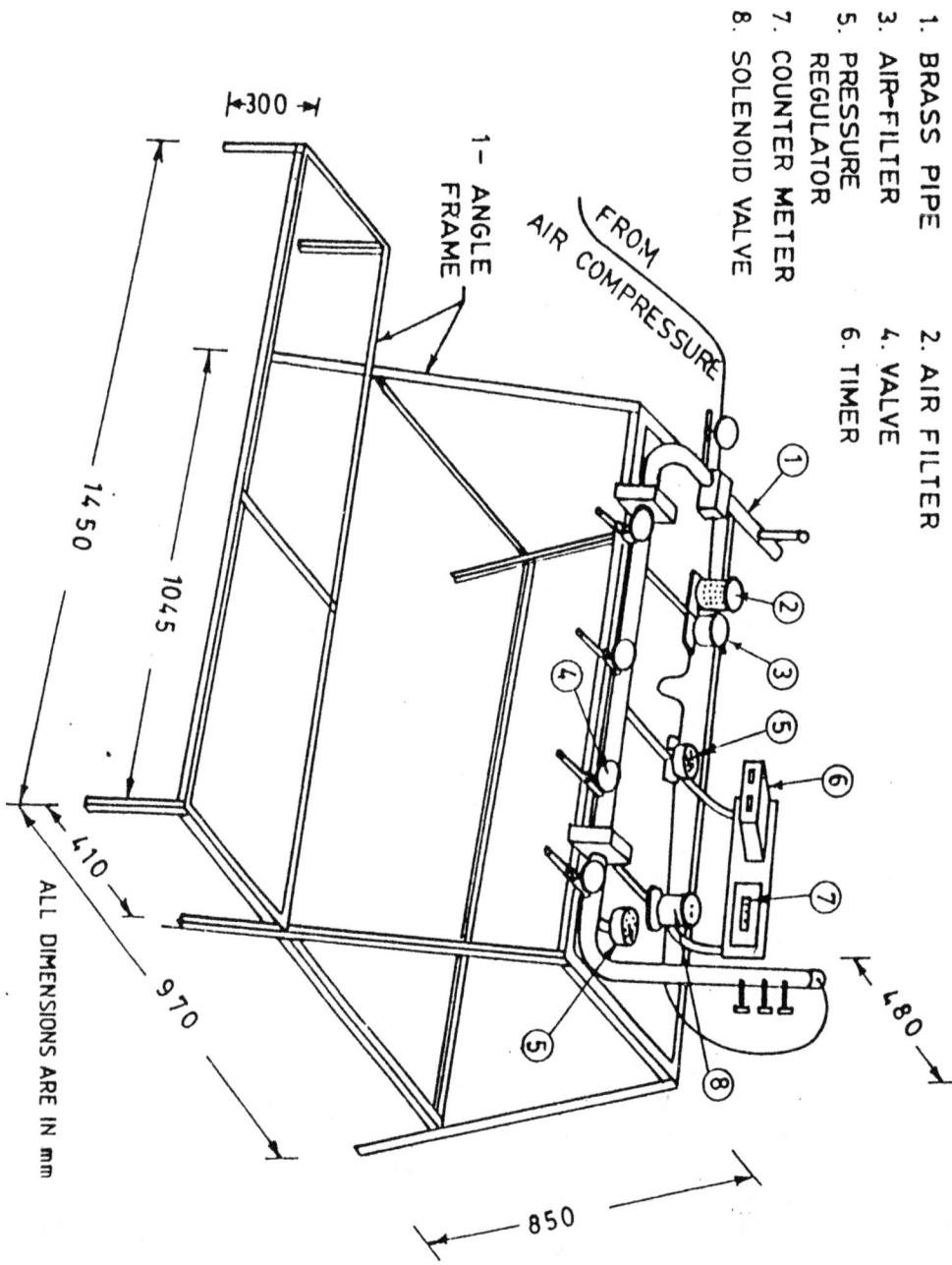

1. BRASS PIPE   2. AIR FILTER
3. AIR-FILTER   4. VALVE
5. PRESSURE   6. TIMER
   REGULATOR
7. COUNTER METER
8. SOLENOID VALVE

1- ANGLE
   FRAME

FROM
AIR COMPRESSURE

ALL DIMENSIONS ARE IN mm

FIG.8.4 TEST RIG FOR FATIGUE TEST ON COMPRESSION SPRAYER

indicate the number of pressure cycles. The tank shall confirm this test if no leakage, crack or deformation of the tank is occurred.

m- *Test for spray lance*

The inlet of spray lance shall be fitted to a hydraulic pump directly or through a delivery hose. The outlet of the lance shall be closed so that no discharge is allowed from the lance. A hydraulic pressure of 1MPs or two and half times of the normal working pressure of the sprayer, whichever is more shall be applied to the lance for a period of 5 minutes. If during this test there is no leakage, crack of burst then the test is confirmed.

4- Performance test for cut-off device

This test shall be conducted with water liquid containing 5% suspension of DDT. The cut-off device shall be rigidly mounted on the test rig as shown in Fig. 8.5 & 8.6 for trigger type and knob type respectively. The test liquid shall be applied to the inlet of cut-off device under s static pressure of 300 ± 30 KPa. The cut-off device shall be operated for 5000 cycles at the speed of 15 cycles per minute and further for 500 cycles at a pressure of 600 ± 60 KPa. The cut-off device is confirmed if it does not leak.

5- Performance test for nozzle

Rate of discharge and spray angle test is conducted on patternator as shown in Fig. 8.5.

The nozzle should be able to provide a rate of discharge ml/min as per chart given below :

S.No.	75 KPa for public health purpose	300 KPa for Agril. purpose
1	150	225
2	200	300
3	225	337
4	300	400
5	337	450
6	400	600
7	450	675
8	600	800
9	675	900
10	900	1200
11	1200	1350
12	1350	1800

13	1800	2400
14	2700	2700
15	3600	3600
16	4500	4500

The rate of discharge also should be within ± 5% of declared value.

To calculate spray angle the nozzle is fitted at the height of 545 mm on patternometer. A long scale is placed on the patternometer to have impression of discharge. Accordingly spray angle is calculated. The value so obtained should not have variation more than + 3° then the declared value in order to confirm this test.

6- Endurance test

The sprayer is operated in accordance with the normal working pressure for a minimum period of 48 hours on the test rig. This period preferably should be in continuous stretches of 6 hours. The sprayer is deemed to have passed this test if no leakage or breakdown is observed during this test.

Norms of important parameters for different type of sprayers as per BIS test code is shown in Appendix-XI.

8.6 Type of tests for dusters

1 General tests

a Material of construction for different components
b- Constructional requirements of hopper, feed control mechanism, agitator and feeder
c- Test for agitator
d- Strap drop test

2 Performance tests

a Air output test
b Dust delivery test
c- Dust throw test
d Leakage test

8.7 Testing method for dusters

1 General test

a Material of Construction for different components.

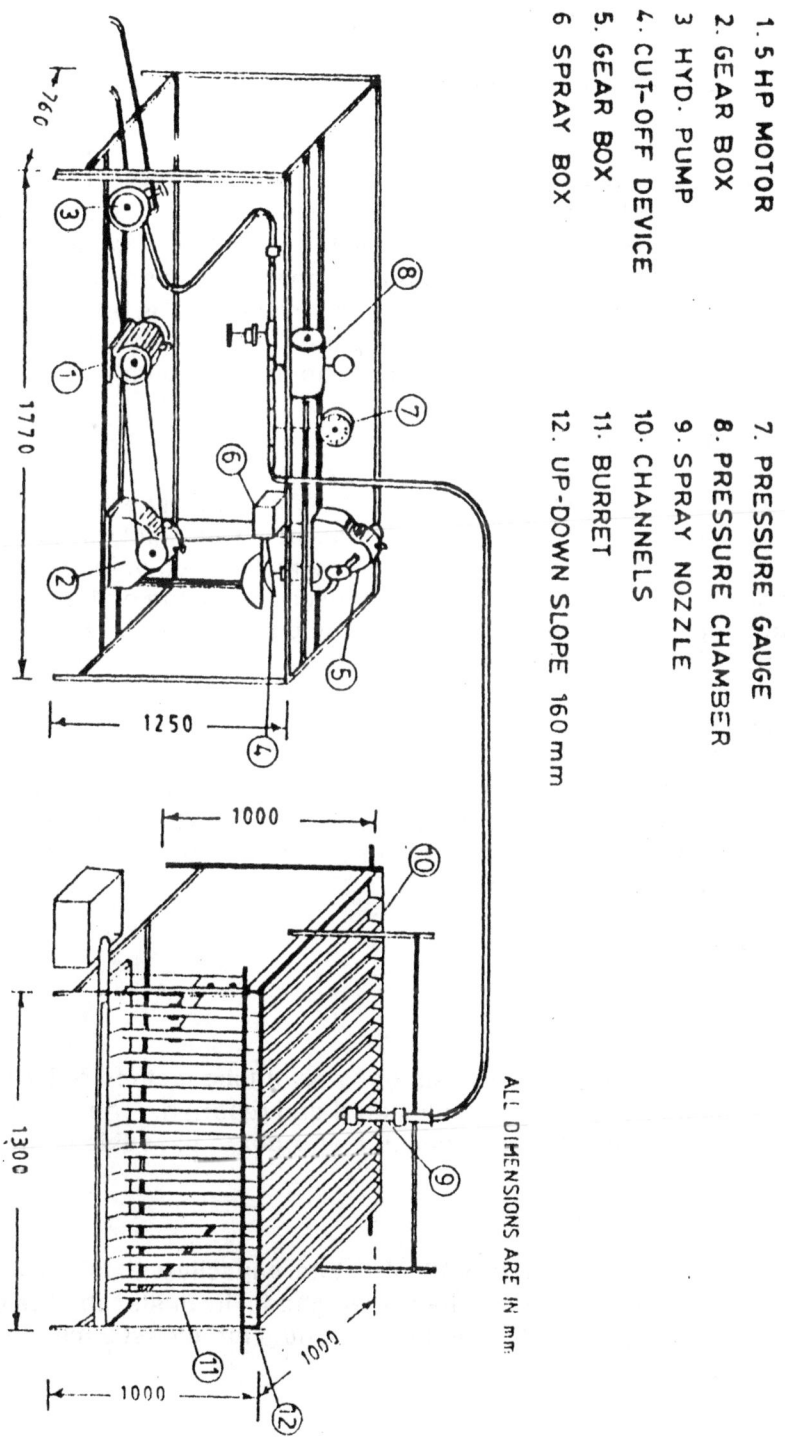

FIG. 8.5 TEST RIG FOR TRIGGER-TYPE CUT-OFF DEVICE

1. 5 HP MOTOR
2. GEAR BOX
3. HYD. PUMP
4. CUT-OFF DEVICE
5. GEAR BOX
6 SPRAY BOX
7. PRESSURE GAUGE
8. PRESSURE CHAMBER
9. SPRAY NOZZLE
10. CHANNELS
11. BURRET
12. UP-DOWN SLOPE 160 mm

PATTERNATOR FOR SPRAY
PATTERN AND SPRAY-ANGLE

ALL DIMENSIONS ARE IN mm

All the metallic parts coming in contact with the pesticide should preferably be of the same material to minimize electrolyte potential deterioration. The material used for different components shall be declared by the manufacturer in the parts catalogue, compared with relevant test code and reported accordingly.

b- Constructional requirement of hopper, feed control mechanism, agitator and feeder:

The hopper shall have a concave shape or conical bottom so that the dust contained in it, moves towards the feeding aperture. On the top of the hopper, a filler hole of atleast 130 mm diameter should be provided. The hole shall be covered with a lid. A feed control mechanism with locking device should be provided to control the flow of dust through the aperture. An agitator either integral with feeder or separately shall be incorporated within the hopper to keep the dust agitated and to avoid the clogging of the aperture for feeding the dust to the aperture smoothly.

c- Test for agitator:

The hopper shall be filled upto 3/4th of its total capacity with talc powder used for insecticidal formulation. The duster shall be fixed rigidly in place and shall be operated continuously at medium discharge rate setting of feed control mechanism as specified by the manufacturer till the discharge at the outlet siezes. The duster shall be considered to confirm the test if the dust remains in the hopper is not more than 0.5% of total mass of dust.

d- Strap drop test:

The hopper shall be filled with talc powder to its total capacity. The duster shall be hung from a solid support by its straps. It shall be lifted to a heights of 30cm and allow to drop. The test shall be repeated for 25times. If no breakage or deformation is occurred, the test is confirmed.

2. Performance test

a- Air out put test :

The air output test is conducted either by pivot tube method or by air flow method. Average value of 5 readings is calculated and reported. The fan should be able to deliver not less than $0.3m^3$ of air/minute in order to pass this test.

b- Dust delivery test:

The hopper shall be filled upto 3/4th of its total capacity with talc powder. The duster with its all working accessories shall be weighed. The duster shall be fixed rigidly and operated uniformly at a speed of 35 rev./minute for atleast 2 minutes. The mass of the duster shall again be determined. The total mass of the dust discharge shall be computed and the rate of discharge/minute determined. The delivery rate at maximum discharge setting should not be less than 150 gm/minute in order to pass this test.

c- Dust throw test:

The hopper shall be filled upto 3/4th of its total capacity with talc powder. Set the duster and delivery pipe at its horizontal position. Operate the duster continuously and uniformly at a speed of 35 rev./minute. Measure the horizontal distance from the outer most point of the delivery pipe and the outer most point where dust falls on the ground. The duster should be able to throw the dust upto a minimum distance of 1 meter in order to confirm this test.

d- Leakage test:

The hopper shall be filled upto3/4th of its total capacity with talc powder. Set the duster at its horizontal position and plug the outlet. Operate the duster at a speed of about 30 rev./minute for two minutes. The duster is considered to confirm this test if no leakage of dust is occurred.

Norms of important parameters for different type of dusters as per BIS test code are shown in Appendix XII.

1. MOTOR 5 HP   2. PUMP
3. MAIN SWITCH   4. STARTER
5. GEAR BOX   6. CRANK
7. RACK   8. PINION
9. CUT-OFF-DEVICE   10. 2" x 2" ANGLE-IRON FRAME

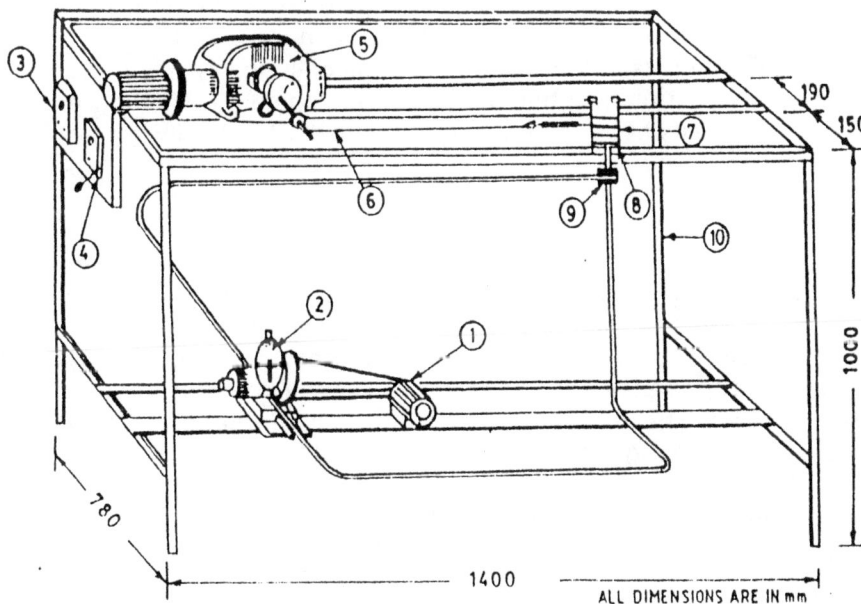

ALL DIMENSIONS ARE IN mm

FIG.8.6 TEST-RIG FOR KNOB-TYPE CUT-OFF-DEVICE

Chapter 9

TESTING AND EVALUATION OF GRAIN COMBINE-HARVESTER-THRESHER

9.1 Introduction

The Grain Combine Harvester Thresher (popularly known as combine) is a machine exclusively to harvest and recover grain from standing plant residue. The combine cuts the crop, feeds the crop to the cylinder, threshes the grain from ear head, separates the grain from straw, cleans the grain and handles the clean grain until it is dumped into a truck or trailer for transport.

9.2 Types of grain combine

Types of combines, generally used in India are:

a) Tractor front mounted

The combine is fitted on the tractor in such a way that it rests on the tractor rear axle as well as on extended portion of rear axle and header unit comes in front of tractor. Tractor's controls are raised to the combine platform with the help of extension levers. The operator sits on the combine platform and operates the machine.

b) Self-propelled combine

The self-propelled combine is powered with industrial type engines of 60 to 150 hp. It is provided with a gearshift or variable - speed drives, such as the hydrostatic drive, to give desired field and road speeds. It is easy to handle and transport from field to field and over the highway. While transportation on highway, the cutter bar platform can be removed from front and hitched behind the combine. The operator of the self-propelled combine sits above and just behind the hydraulically controlled platform.

9.3 Combine systems

Five major functions are performed during the harvesting operation with a combine. These may be classified as (1) cutting the crop and feeding to the cylinder, (2) threshing the grain from the ear head, (3) separating the grain from the straw, (4) cleaning the grain, and (5) handling the grain after threshing. These functions are automatically performed as the material is moved through different systems of combine harvester. The schematic diagram for self propelled and tractor mounted combines is shown in Fig. 9.1 and 9.2 respectively.

The standing crop is guided to the cutter bar platform by the reel. Dividers divide

1.KNIFE 2.REEL 3.CROP DIVIDER 4.CONVEYOR AUGER 5.FEEDING FINGERS
6.FEEDING CONVEYOR 7.THRESHING CYLINDER 8.STRAW GUIDE DRUM 9.STRAW WALKER
10.CATCH 11.FLAP SIEVE 12.RAKE 13.EAR RETURN 14.BOTTOM SIEVE 15.BLOWER FAN
16.SLIDE 17.COLLECTOR BOTTOM 18.GRAIN AUGER 19 GRAIN ELEVATOR 20.TANK FILLING
AUGER 21.GRAIN TANK 22.TANK AUGER 23.EAR AUGER 24.CONCAVE

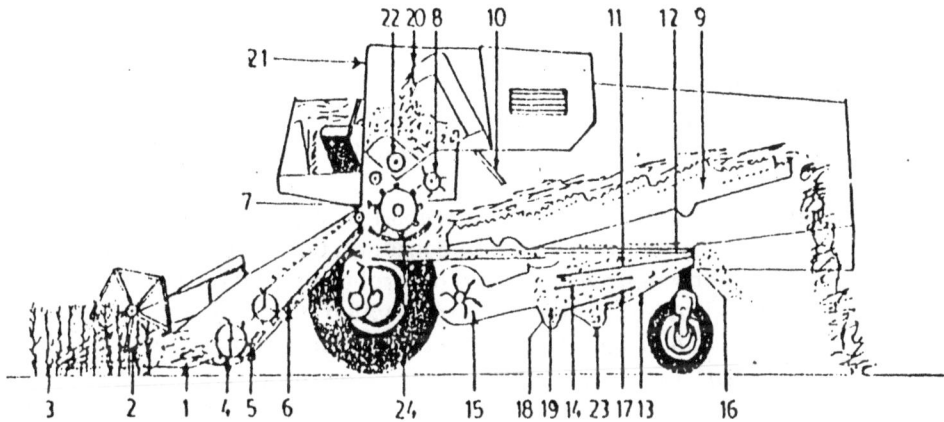

FIG.9.1 SELF-PROPELLED COMBINE HARVESTER

1.REEL 2.CUTTER BAR 3.AUGER 4.FEEDING CONVEYOR 5.CONCAVE
6.THRESHING DRUM 7.BEATER 8.STRAW WALKER 9.SIEVE
10.GRAIN TANK 11.TAILINGS ELEVATOR 12.DISCHARGE AUGER
13.GRAIN ELEVATOR

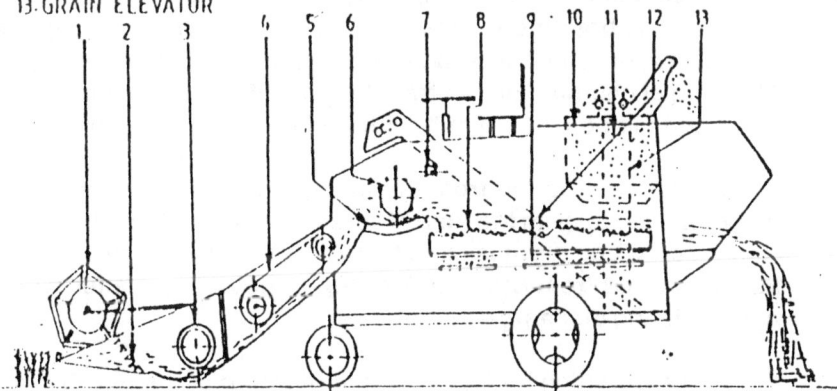

FIG.9.2 TRACTOR OPERATED COMBINE HARVESTER

the crop which has to come to the machine and the rest standing in the field to be slightly pushed apart so that it is not damaged by the machine. The crop is harvested by the cutter bar and the cut crop falls into the pan of the cutter bar. It has rotating auger which brings the cut crop to the centre of auger. The centre of auger is fitted with retractable fingers which conveys the crop to feeder channel. The feeding chain fitted with slats, running in feeder channel conveys the crop to the threshing unit. The crop is threshed by threshing drum and concave. To suit various crops and their threshing characteristics, drum speed and the gap between concave and drum can easily be adjusted.

A stone trap is provided in front of threshing drum and concave so as to catch the stone coming alongwith crop. Stone being heavy in weight drops in the stone trap and remains there to be taken out during the servicing of the machine.

The threshed straw is brought to the straw walker by guide drum. The mixture of grain and short straw (chaff) returns to step bed through bottom pan of straw walker. A baffle plate is placed above straw walker to control the straw to the straw walker by allowing the straw to fall at front end of the straw walker thereby making it to travel complete length of straw walker so that left over grains are dropped out to the pan of the straw walker. The straw is thrown out from discharge hood. The mixture of grain and chaff dropped from the grates of the concave and also through the straw walker fall on the step bed. Due to vibrating action of step bed, grains are separated from the chaff and short straw and carried to the chaffer sieve and cleaning sieves. When the grains, chaff and short straw are being dropped on the chaffer sieve, air is directed from the blower which blows the chaff and short straw out of combine. The unthreshed ears drop through chaffer rack to the ear return pan. Under the adjustable sieve another fixed hole sieve is placed. The inclination of this sieve is adjustable. It is also replaceable and can be changed to suit the type of crop.

Air from variable speed blower is arranged in such a way that it passes under chaffer rack as well as both the sieves. The air current blows out light weight dust particles, short straw, green trash and other impurities out of the combine. A slide is mounted at the end of cleaning box to trap the grain if any being carried away by the air current. The slide is adjustable in height and can be locked in any position.

The clean grains are dropped from the fixed hole sieve to the inclined lead and carried to the grain elevator with the help of screw conveyor. At the top end of the grain elevator another screw conveyor is fixed to convey the clean grain to grain tank. The grain can be discharged from grain tank to the transporting trailer with the help of discharge conveyor and discharge auger.

The unthreshed ears which had fallen at the rear part of the chaffer rack find their way on the inclined bed and carried by screw conveyor into the threshing drum for re-threshing.

9.4 Test items

A- Laboratory tests

1- Specification checking
2- Pre-test checking
3- Material analysis
4- Engine performance
5- Noise level
6- Vibration level
7- Visibility
8- Brake performance
9- Air cleaner oil pull over
10- Centre of gravity
11- Turning space and turning circle
12- Endurance
13- Power drop
14- Wear and tear

B- Field tests

1- Rate of work
2- Quality of work
        a- Grain losses
        b- Rubbish content
        c- Grain damage
        d- Threshing efficiency
        e- Cleaning efficiency
3- Output of straw and grain
4- Fuel consumption
5- Night trial observations
6- Ease of operation and handling
7- Safety provisions
8- Soundness of construction
9- Labour requirements
10- Handling characteristics

9.5 Procedure for laboratory testing

1- Specification checking

The specifications of the combine given by the applicant shall be checked. compared and reported by the testing authority. While checking various dimensions, following items need to be considered:

a- Name and address of the manufacturer
b- Country of origin
c- Make
d- Model
e- Sl.No.
f- Prime mover and its details
g- Chassis
h- Wheels with their details
i- Reel assembly with details
j- Cutter bar assembly details
k- Combining system details
l- Threshing Drum assembly details
m- Straw walker details
n- Cleaning and chaffer sieves details
o- Grain conveying mechanism details
p- Speeds
q- Lubrication of the combine harvester and lubrication schedule
r- Fuels and lubricants to be used
s- Light arrangements
t- Safety devices and special safety feature

2- Pre-test check

The combine is installed on plain surface and examined for pre-test checks with particular attention to:

i- Bearings
ii- Drives and other moving parts
iii- Correctness of various system
iv- Proper adjustments of various assemblies
v- Tightness of bolts and nuts

A check shall also be made to ensure that all accessories which are normally delivered with the machine have been provided.

3- Material analysis

The hardness and chemical analysis of critical components like knife section. raspbar, peg tooth, ledger plate and knife guards shall be made and reported.

4- Engine performance

This test is conducted only in case of self-propelled combine.

General requirements

Run-in of the engine

a- The engine of the combine will be run-in in accordance with the procedure laid down in the instruction manual. During this period drives and bearings will be checked for overheating, alignment. excessive vibration. etc.

In the absence of specific instructions about run-in, the testing station shall carry out run in as under:-

    i- 5 hours at quarter load at speed specified by the manufacturer for continuous operation

    ii- 5 hours at half load at speed specified by the manufacturer for continuous operation

    iii- 5 hours at three quarters of load at speed specified by the manufacturer for continuous operation

    iv- 5 hours at full load at speed specified by the manufacturer for continuous operation.

b- The torque and power values in the test report should be obtained from the dynamometer without correction for losses in engine auxiliaries.

c- In all tests. the shaft connecting the power take-off to the dynamometer should have the minimum angularity.

d- During maximum power test, ambient temperature should be between 25 degree C and 35 degree C and atmospheric pressure should not be less than 96.6 kpa. Atmospheric pressure and ambient temperature should be noted during all the test readings for power and average reported.

e- For all test. all accessories shall be in position on the engine. All attachments necessary for running of the engine shall be attached in the same position.

f- The various tests shall be carried out continuously, the governor controllever be placed in the position recommended by the engine manufacturer for obtaining continuous maximum power.

g- In addition to the performance measurements, the following should also be noted:

1. i) The temperature of fuel measured on fuel measuring apparatus.
   ii) The temperature of engine lubricating oil measured near the drain plug.
   iii) The exhaust gas temperature measured near the final junction of exhaust manifolds.

2. The coolant temperature shall be measured at the outlet of coolant flow line between cylinder block or cylinder head and thermostat. In case of air cooled engine, the temperature of air emitting from cylinder block shall be measured at 2 points at a maximum distance of 4 cm from cylinder fins.

3. The air temperature measurements shall be made approximately 2m in front of the air inlet and approximately 1.5m above the ground to ascertain the effect of exhaust gases on the air intake.

For engines fitted with a blowing device, air temperatures shall be taken at approx. 2m behind the engine and approx. 1.5m above the ground and at the engine air intake.

4. Smoke density: Smoke density exceeding 5 bosch number is unacceptable.

5. Pressure of the engine lubricant.

6. Relative humidity of air.

7. The specific fuel consumption figures in the test report should be given as gms.of fuel per horsepower per hour. The specific fuel consumption at all test shall be noted but the specific fuel consumption at corrected maximum hp shall be considered as base figure.

h- To obtain hourly consumption the volume and work performed per unit volume of fuel conversion of weight to units of volume should be made using the density value at 20⌐C.

i- When consumption is measured by volume, the specific fuel consumption should be calculated using the density corresponding to the appropriate fuel temperature corrected to 20⌐C.

j- Oil consumption: The observation for this test should be taken from 5 hrs. test (high ambient) and report of oil consumption be made as gms/b hp/hr.

The comments regarding specific fuel consumption for test on combine engines may be assessed on the following:-

Rating	Sp.fuel consumption(g/b hp/hr)
Normal	170 to 200
Slightly· high	200 to 225
High	Above 225

Procedure of engine tests

a- Varying speed test

The varying speed test shall be conducted under natural ambient conditions (temp. range: 30 ± 5 degree C). It shall be repeated under high ambient condition (temp. range = 45 ± 2 degree C) to assess engines performance under tropical climatic conditions.

b- Test at maximum power

The engine shall be operated for a period of two hours after it has warmed up, for power to become stabilized. The test shall be carried out under natural ambient conditions. A minimum of six readings at 20 minutes interval of time shall be made during two hours test period. Temperature, pressure and other observations will be recorded simultaneously.

The maximum power quoted in the report should be the average of the readings made during two hour period. If the power variation exceeds ± 2 per cent from the average, the test should be repeated. If the variation continues to exceed ± 2% it should be mentioned in the report.

c- Tests at varying load

In the zone controlled by the governor, the torque, speed and hourly fuel consumption shall be noted as a function of power. In addition the no-load engine speed shall be recorded. The data required to complete the varying load test shall be recorded in the following loads and sequence of measurements. Each load shall be maintained for a duration of 20 minutes.

> i- 85 percent of load obtained at maximum power
> ii- On minimum load
> iii- 50% of the load defined in (i)
> iv) A load corresponding to maximum power
> v) 25% of the load defined in (i)
> vi) 75% of the load defined in (i)

d- Five hours high ambient rating test

The engine shall be run at 90% of maximum output continuously for 4 hours under high ambient temp. (45±2 C). The engine shall be run at a load corresponding to maximum power for a period of 1 hour running after 4 hours of continuous high ambient test at 90% load corresponding to maximum power.

The readings of speed, power, torque, temp., pressure and fuel consumption shall be taken after every half an hour during 4 hours period and after every 15 minutes during 5th hour run. Coolant and lub. oil consumption shall be noted and reported in the test report.

Presentation of curves

The test report should include the following curves made for full range of engine speeds available under natural ambient and high ambient conditions:-

i- Power as a function of engine speed.

ii- Equivalent crank shaft torque as a function of engine speed.

iii- Hourly fuel consumption as a function of engine speed.

iv- Specific fuel consumption as a function of power.

v- Specific fuel consumption as a function of speed.

5. Noise level measurement

A good quality sound level meter shall be used. Measurement shall be made on the combine with no-load, in sufficiently silent and open zone. For example, this zone may be an open space of 50 m radius, of which the central part of atleast 20m radius shall be practically level and be firm with a smooth surface. A concrete or similar surface may be preferred. Measurements shall be made in good weather conditions with little or no wind. Any extraneous noise occurring during the reading, which is not connected shall not be taken into consideration.

Test method

The combine shall be operated at the recommended speed for field work. All the mechanism in the combine shall be in working position. The cutter bar height shall be kept at 15 cm above the ground level. Noise measurements at a height of 1.2 ± 0.1m from ground level shall be made at the following four points when the microphone is facing the combine.

a- In front of combine 3 m from the fore-most point of the combine

b- In the rear of the combine 3 m from the rear most point of the combine

c- Both the sides of the combine 2 meters away from the wheels or outer end

d- Measurement shall be considered valid if the difference between the 2 consecutive measurement from the same side of the combine is not greater than 2 decibles. The value shall be that corresponding to the highest sound level.

Noise at the driver's ear level

a- Measurement of sound frequency spectrum shall be carried out by using a frequency analyzer fitted with octave filters.

b- The microphone shall be positioned atleast 4 cm and not more than 7 cm to the side of the driver's fore-head.

c- The report shall state whether the test was carried out with or without a cab on the combine.

d- The highest noise level and any unusual characteristics giving noise to the operator shall be stated in the report.

6. Vibration level test

The amplitude of mechanical vibration of various sub-assemblies and components of the combine shall be measured by using precision vibration meter. The specifications of the instrument used shall be included in the report. The observations shall be recorded by parking the combine on a level concrete surface and operating the combine at the speed recommended for field work. The cutter bar shall be kept at 15 cm above ground level. The readings of vibration may be taken on important assemblies/sub-assemblies/ components of the combine. The maximum displacement of vibration shall be measured in horizontal and vertical position and represented in microns for horizontal and vertical displacement.

7 Visibility test

A- Visibility from driver's seat

The combine shall be parked on a level horizontal surface. The height of vision during test shall be maintained at 760 mm on a vertical plane from the centre of operator's seat. The line separating the area of visibility and that obstructed by the combine shall be marked on the surface. A graphical representation of the area marked on the surface shall represent the area visible from operator's seat and the area not visible from the operator's seat.

The result of the visibility test shall be graphically represented in the test report.

8- Brake test

The brake test shall be carried out on an artificial horizontal track giving a good grip for tyres. The surface shall be dry and clean. Before commencing the tests it shall be verified that the settings and condition of the brake components conform to the manufacturer's specifications. The tyre pressure shall be as recommended by the manufacturer for the field work. The pressometerm shall be fitted on brake pedal and read out device placed suitably on the combine.

138

The combine shall be declutched and the brakes applied successively increasing force on the pedal, until that giving the shortest stopping distance is found. The following shall be recorded.

a- Deceleration as measured by a 'maximum decelerometer'
b- Stopping distance
c- Force exerted on the brake pedal, and,
d- Braking efficiency

The details of the instrument used for brake tests shall be indicated in the test report.

9- Air cleaner oil pull over test

The mass of oil in the air cleaner assembly (with recommended grade of oil) shall be 5% in excess than the marked level. Then engine is run for 15 mts. followed by sudden acceleration and deceleration made after every 30 secs for 15 mts. The oil pull over in grams and percent of oil pull over shall be calculated for 5 different position of combine i.e. horizontal, tilted 10" forward. 10" backward, 10" left side and 10" right side.

10- Centre of gravity

Position of centre of gravity shall be determined when the combine is fitted with all standard accessories, all the liquid reservoirs and grain tank full and operator replaced by 75 kg mass on the seat. The header assembly shall be in full raised position and reel in its most forward position.

11- Turning space and turning circle

Turning space is defined as the diameter of the circle formed by the outer most point of the combine when it takes shortest turn.

Turning circle or turning radius is defined as the distance from the turning centre to the centre of ground contact of the wheel forming the largest circle when the combine takes shortest turn.

Turning space and turning circle readings shall be taken for left and right turn when the combine is running at speed less than 4 km/hr.

12- Endurance test

Endurance test of 250 hours shall be carried out on the combine after the field work has been completed. The endurance test shall consist of running the engine at test recommended speed and the combine may run empty. After the completion of the endurance test, general condition of the combine shall be checked, with special attention to welded joints and components made of sheet metal. This is optional test to check durability of the combine.

13- Power drop test

A two hours maximum power test shall be conducted after the end of field test and endurance test to determine drop in the power. Except the normal adjustments specified by the manufacturer in the printed service manual, no special adjustments or change of parts shall be allowed. The comparative drop in horsepower should not be more than 5%, and comments shall be given accordingly in test report.

14- Wear test

The engine (in case of self-propelled combine) parts, clutch, transmission system, starter and alternator etc. shall be dismantled and assessed for rate of wear (percentage basis). The assessment will be made keeping in view the initial average values (when the engine is new) as also the maximum permissible wear as specified by the manufacturer. The rate of wear in respect of rasp-bar, peg tooth, concave shall be computed in terms of percent mass basis on the basis of original mass before test and final mass after test.

9.5 Procedure for field testing

The manufacturer's representative should demonstrate the operation to test team in actual field conditions (crop). The officer Incharge of the test should take note of any feature requiring particular attention. The machine should be driven by an experienced operator. Before starting the test, the machine should be adjusted following the instruction manual in order to obtain the highest possible performance. During test run no alteration in adjustment or speed is allowed. However, the combine may be readjusted between successive test runs in order to have maximum possible performance.

Field and crop condition

The combine shall be operated for minimum of 200 hrs preferably in the following crops:

1-	Wheat	100 hr
2-	Paddy	50 hr
3-	Any other crop	50 hr

If any other crop is not recommended then the 50 hours should be adjusted in paddy crop. The test crop shall be ripe (ready for harvesting), and standing angle should be more than 60 degrees. Other crop conditions like variety, height of plant, length of ear head, tillings, no.of grains in ear head plant density and moisture content of grain and straw etc. shall be recorded for each test.

Field conditions, like topography, type of soil, weed intensity, size and shape of field etc. should also be recorded together with important observations on crop conditions and plant parameters.

9.7 Important field terms

1-Collectable loss: The unthreshed ear heads in main grain outlet or grain tank; threshed, unthreshed and damaged grains from the secondary cleaning or grading unit.

2- Non-collectable loss: The header loss, shoe loss, rack loss and secondary blower loss.

3- Header loss: The loss of grains and ear heads being shed and left over on the ground as a result of operation of cutter bar and header unit.

4- Preharvest loss: The loss of grain or ear heads from the standing crop prior to the operation of combine in the field.

5- Processing loss: The damaging of the grain, unthreshing of grains, loss of threshed (damaged or undamaged) grain and unthreshed grains after completion of threshing, separating and cleaning operations.

6- Rack loss: The threshed grains passing out in the straw.

7- Shoe loss: The threshed grains blown or carried out with the chaff.

8-Cleaning efficiency: Clean grains percent in the total grain obtained from the main grain outlet expressed in percentage by mass.

9 Field efficiency: The quotient of effective field capacity and theoretical field capacity expressed in percent.

10- Threshing efficiency: Threshed grains from all the outlet of the combine with respect to grain output in tank expressed in percentage by mass.

11- Grain output: The mass of the grain mixture delivered by the combine per unit of time.

12- Straw output: The mass of the straw and chaff not including grain losses delivered by the combine per unit of time.

13- Unloading rate: The volume of grain unloaded per unit time expressed in cubic centimetres per second.

14- Combine capacity: The maximum sustained total feed rate measured in kilogram per second or tonnes per hour at which the rack loss and shoe loss shall be within the acceptable limits while the combine is operated at rated speed on level ground without chocking of threshing, separating, cleaning and grain conveying mechanism& without stalling of prime mover.

9.8 Material required for measurement and sampling during field test

i) Canvass sheets or cloth sheets for collection of straw walker and sieve sample during 20 m test run.

ii) Platform balance or tripod balance for weighing straw and chaff samples.

iii) Sighting poles (atleast 4 Nos.)

iv) Whistle or flag for giving signal at the beginning and end of measuremen test run.

v) Stop watch for recording running time of combine harvester during test run.

vi) Bags for collection of grain and mixture sample.

vii) Measuring tape and rules for measurement of test run, test plot, crop and plant parameters.

viii) Pan balance for weighing sample for analysis and moisture.

ix) Containers for grain sample and moisture sample.

Procedure for test (Fig.9.3)

i) When the combine harvester is ready for test, it should be topped up and be operated in the plot at uniform speed.

ii) A test run of 20m is selected from the test plot and marked with sighting poles.

iii) Preharvest losses at three different places randomly selected having an area of 1m x half the cutter bar width be determined in kg/ha. A rectangular frame of appropriate size be used for marking the area.

iv) Two roll of canvass or cloth having 30m length and one and half times width then that of straw and sieve out-let are rolled over on the specially attached rollers behind the combine to collect straw and sieve sample.

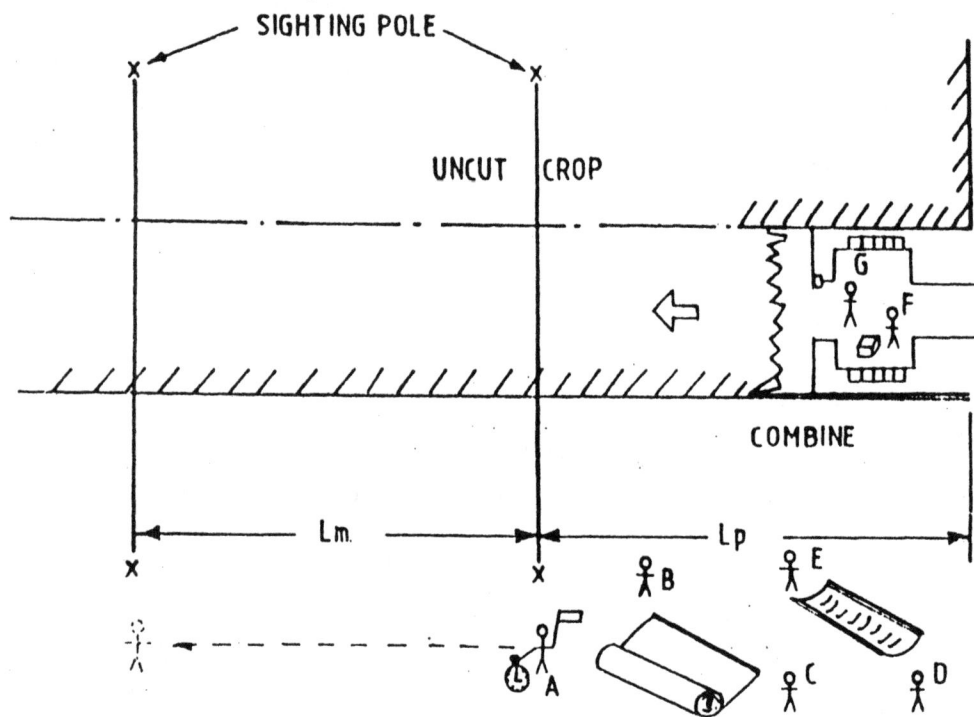

WHERE

    Lp : LENGTH OF PRELIMINARY RUN

    Lm : LENGTH OF MEASUREMENT (TEST) RUN

    A : OBSERVER FOR SIGNAL

    B,C : OBSERVER FOR COLLECTION OF STRAW WALKER SAMPLE

    D,E : OBSERVER FOR COLLECTION OF SIEVE SAMPLE

    F : OBSERVER FOR SAMPLE OF GRAIN OUTLET

    G : COMBINE OPERATOR

FIG. 9.3 ARRANGEMENT FOR FIELD TESTING OF COMBINE HARVESTER

v) Arrangement for collection of sample shall be made as per Fig. 9.3.

vi) The combine, when enters the test run, signal shall be given to record the time taken to cover the 20 m test run and to collect different samples.

vii) The header loss shall be determined from that portion from where pre-harvest losses were recorded and during test it was protected by cloth sheet.

viii) All the data shall be recorded as shown in Appendix-XIII.

9.9 Observations to be recorded during and after test

i)   Area covered
ii) Time of operation
iii) Operating speed
iv) Time for any stoppage
v)   Time loss in turning (this may be recorded for atleast one hour operation)
vi) Average working width
vii) Time required to discharge the grain from grain tank
viii) No load and on load engine speed
ix) Maximum temperature of engine oil, coolant, transmission oil, hydraulic oil and ambient
x)   Fuel consumed (l/h and l/ha)
xi) Lubricating oil consumed
xii) Coolant (water) consumed
xiii) Average forward speed (km/h)

9.10 Sample analysis

Three samples of 100gm each from the main grain outlet shall be taken and analyzed for threshed, unthreshed, broken and rubbish content. Similarly complete sample for the test run from straw walker and sieve shall be analyzed. The complete analysis shall be done as per Appendix-XIV.

9.11 Data analysis

The data obtained during field test as available in Appendix-XIII and sample analysis as shown in Appendix-XIV will be used for analysis and presented as per Appendix-XV. Then the following results can be obtained.

1- Rate of work: It can be represented in terms of :

      a-   Area covered (ha/h)
      b-   Net grain output (Kg/h, Kg/ha)

c- Grain throughout (Kg/h, Kg/ha) :
It consists of net grain output, header loss, threshing cylinder loss, rack loss and shoe loss.

d- Straw output (Kg/h, Kg/ha)

2- Quality of work

a- Grain losses: The collectable loss and non-collectable loss are calculated as per Appendix- XV.

b- Rubbish content: Any thing other than grain is called rubbish content and is determined on the basis of 100 gm representative sample of grain from main grain outlet.

c- Grain damage: A representative of 100 gm sample of grain from grain outlet is used to determine grain damage.

d- Threshing efficiency: It is expressed in percent of threshed grain with respect to grain output as shown in Appendix -XV.

e- Cleaning efficiency: It is expressed in percent of clean grain with respect to total grain obtained from main grain outlet as shown in Appendix-XV.

3- Output of straw and grain

Grain output and straw output in kg/h and kg/hA can be calculated as per Appendix-XV.

4- Fuel consumption

Fuel consumed during each test shall be computed by topping up at the start of test and at the finish of test and expressed in terms l/hr and l/ha. After complete testing the range of fuel consumption shall be reported in the test report.

5- Night trial observations

A night trial of minimum two hours each shall be conducted in wheat and paddy crop to assess the intensity and suitability of the lighting equipments for the night work.

6- Ease of operation and handling

Observations shall be made on skill and intensity of effort required to operate various controls of the machine. Adequacy and accessibility of controls and visibility of the header and instrumentation shall also be recorded. The note on operator's working condition, the ease of setting adjustment, routine maintenance and other similar features shall also be made.

7- Safety provisions

The safety devices such as slip clutches, shear pin, signal horns, indicator lights etc. provided for various systems shall be assessed for their effectiveness. Provision of stone trap, spark arrester and automatic intermitant horn while reversing of combine shall be checked and reported accordingly.

8- Soundness of construction

Observations shall be recorded of those features which adversely affect the operation and efficiency of combine in the field. The test will be completed with a thorough check-over of the test combine, with dismantling wherever necessary to note any weakness which has shown up since the pre-test check, and which may effect the life of the combine. Modifications which could bring about improvement in the quality of work shall be noted. Any weakness which has been seen since the pre-test check or excessive wear will be reported together with details of breakdowns, defects and replacement of parts.

9- Labour requirement

The labour required for operating the combine and for routine servicing and adjustments shall be recorded and reported.

10- Handling characteristics

Detailed observations shall be recorded regarding skill and intensity of efforts required to operate the machine. Adequacy of visibility will be measured. Location of seat, steering brake, controls etc. shall be noted from handling point of view. Accessibility and ease of adjustment and problem in routine maintenance will be noted and reported.

9.12 Summary of performance parameters required for analysis during field test

       a- Duration of test
       b- Forward speed
       c- Rate of work (ha/h)
       d- Net grain output (Kg/h)
       e- Fuel consumption (l/h and l/ha)
       f- Preharvest losses (Kg/ha)
       g- Cutter-bar losses (%)
       h- Total straw walker losses (%)
       i- Total sieve losses (%)
       j- Total collectable losses (%)
       k- Total non-collectable losses (%)

l- Grain breakage or damage (%)
m- Threshing efficiency (%)
n- Cleaning efficiency (%)
o- Moisture content of grain
p- Moisture content of straw

Example for Calculation of grain losses and other parameters

During combine testing, the following data was recorded:

Avg. time of 20 m length	= 22.98 Sec.
Avg. width of cut	= 3.98 m
Weight of grain sample	= 29000 g
Analysis of 100 gm sample	

Healthy threshed	= 92.72 g
Unthreshed	= 0.617 g
Broken	= 1.43 g
Rubbish	= 5.233 g

Calculation value of total grain sample

Healthy threshed grain	= 26888.8 g
Unthreshed grain	= 178.93 g
Broken grain	= 414.7 g
Rubbish	= 1517.57 g

Analysis of straw sample

Weight of straw sample	= 12.2 kg.
Healthy threshed grain	= 24.8 g
Unthreshed grain	= 6.4 g
Broken grain	= 0.9 g
Straw	= 12167.9 g

Analysis of chaff sample

Weight of chaff sample	= 7.8 kg.
Healthy threshed grain	= 42.7 g
Unthreshed grain	= 2.4 g
Broken grain	= 5.1 g
Chaf & Bhusa	= 7749.8 g
Weight of preharvest	= 3.68 g

losses in 1 m x half of cutter bar width
Wt. of cutter bar losses = 8.4 g in 1 m x half of cutter bar width

Calculations:

Area covered in 20 m run = 3.98 x 20 = 79.6 m²

Critical rate of work at the time of sample collection $= \dfrac{\text{Area covered x 0.36}}{}$

$$= \dfrac{79.6 \times 0.36}{22.98}$$

$$= 1.25 \text{ ha/h}$$

Total threshed grain from all sources (Healthy + broken)
= 26888.8 + 24.8 + 42.7 + 414.7 + 0.9 + 5.1 = 27377 g

Rubbish = 12167.9 + 7749.8 = 19917.7 g

Total grain from all sources = grain sample + grain from straw & chaff
sample
= 29000+32.1 + 50.2
= 29082.3 g

Preharvest losses (kg/ha) =

$$\dfrac{\text{Wt.of preharvest grain in 1 m x half of the cutter bar width x 10}}{\text{half the cutter bar width (m)}}$$

$$= \dfrac{3.68 \times 10}{2.16} = 17.04 \text{ kg/ha}$$

$$\text{Grain output (kg/h)} = \dfrac{3.6 \times \text{Wt. of grain sample}}{\text{Avg. time for 20 m length}}$$

$$= \dfrac{3.6 \times 29000}{22.98} = 4543.08$$

$$\text{Grain output (kg/ha)} = \dfrac{10 \times \text{Wt. of grain sample}}{\text{Area covered in 20 m run}}$$

$$= \frac{10 \times 29000}{79.6} = 3643.22$$

Cutter bar losses (kg/ha)

$$= \frac{10 \times \text{Wt. of cutter bar losses in each of 1 m} \times \text{half cutter bar width}}{\text{half of cutter bar width}}$$

$$= \frac{10 \times 8.4}{2.16} = 38.89$$

$$\text{Grain throughput (kg/ha)} = \frac{\text{Total grain} \times 10}{\text{Area covered in 20 m}} + \text{cutter bar losses}$$

$$= \frac{29082.3 \times 10}{79.6} + 38.89 = 3692.45$$

Grain throughput (kg/h) = Grain throughput (kg/ha) x Rate of work (ha/h)

= 3692.45 x 1.25 = 4615.56

$$\text{Straw output (kg/ha)} = \frac{\text{Wt. of straw} \times 10}{\text{Area covered in 20 m}}$$

$$= \frac{19917.7 \times 10}{79.6} = 2502.22$$

Straw output (kg/h) = Straw output (kg/ha) x Rate of work (ha/h)

= 2502.22 x 1.25 = 3127.77

$$\text{Crop throughput (t/h)} = \frac{\text{Grain throughput (kg/h)}}{1000} + \frac{\text{Straw output (kg/h)}}{1000}$$

$$= \frac{4615.56}{1000} + \frac{3127.77}{1000} = 7.74$$

Losses due to combine (%)

A. Collectable

i) Unthreshed from main outlet $= \dfrac{178.93 \times 1000}{79.6 \times 3692.45} = 0.61$

ii) Broklen from main outlet $= \dfrac{414.7 \times 1000}{79.6 \times 3692.45} = 1.41$

B. Non-collectable

i) Header loss $= \dfrac{38.89 \times 100}{3692.45} = 1.05$

ii) Straw walker losses :

a) Threshed $= \dfrac{24.8 \times 1000}{79.6 \times 3692.45} = 0.08$

b) Unthreshed $= \dfrac{6.4 \times 1000}{79.6 \times 3692.45} = 0.02$

c) Broken $= \dfrac{0.9 \times 1000}{79.6 \times 3692.45} = 0.003$

Total straw walker losses (%) = 0.103

iii) Sieve losses

a) Threshed $= \dfrac{42.7 \times 1000}{79.6 \times 3692.45} = 0.15$

b) Unthreshed $= \dfrac{2.4 \times 1000}{79.6 \times 3692.45} = 0.01$

$$\text{c) Broken} = \frac{5.1 \times 1000}{79.6 \times 3692.45} = 0.02$$

Total sieve losses (%)   = 0.18

$$\text{Threshing efficiency (\%)} = \frac{\text{Total threshed grain}}{\text{Total grain}} \times 100$$

$$= \frac{27377}{29082.3} \times 100 = 94.14$$

$$\text{Cleaning efficiency (\%)} = \frac{\text{Healthy threshed grain in main outlet} \times 100}{\text{Wt. of grain sample}}$$

$$= \frac{26888.8}{29000} \times 100 = 92.72$$

Total collectable losses (%)   = 2.02

Total non-collectable losses (%)   = 1.33

Total losses (%)   = 3.35

Chapter 10

TESTING AND EVALUATION OF POWER THRESHER

10.1 Introduction

Threshing is one of the most important crop processing operations to separate the grains from the earheads or the plants and to prepare it for the market. Various types of threshers are available in the market like spike-tooth type. chaff-cutter type etc. meant for threshing of specific crop. Recent trend is towards popularization of multi-crop threshers to take care of the crops grown under the changing crop rotation practices. Although this type of thresher can thrash a variety of crops like wheat Barley. Sorghum, Soybean, maize etc. but it is unsuitable for paddy. Axial flow threshers introduced by IRRI have been modified to serve as multi crop thresher which can handle paddy also. The detail of generally used three types of threshers is shown in Fig. 10.1 to 10.3.

10.2 History of development of threshing devices

Long ago. the threshing was done by manual flailing operation, later replaced by treading under animal hooves. These were time consuming methods involving drudgery, exposure of the crop to natural hazards of rain and fire and also loss of grains by animals. birds and insects etc.

The threshing of wheat crop is comparatively difficult as it involves detaching the grains from the earhead as well as breaking up of the straw into 'Bhusa' of an acceptable quality. The threshing of wheat by treading under animal hooves accomplished both the detaching of grains as well as making of bhusa. The use of wooden and steel rollers pulled by a pair of animals improved the output. The introduction of Olpad thresher was a break through which enabled a four-fold increase in output from a pair of bullocks. The 'olpad' thresher had its origin at a place named 'olpad' in south Gujarat. This thresher has three gangs of notched discs mounted on a frame. The farmer can sit on it while driving the animals over the crop.

The use of mechanical power for threshing started with use of chaff cutters in Punjab. It was used to chop up the wheat crop which also partially threshed the heads. The crop was later trampled to complete the threshing in a comparatively shorter period. A modified chaff cutter with the enclosed flywheel having a corrugated periphery and a concave underneath eliminated the use of animal treading and thus paved the way for development of wheat power thresher. This type of thresher was named as "syndicator". In order to obtain clean grain the threshed crop was to be winnowed in a separate operation. M/s Friends Own Foundry of Ludhiana under the guidance of Sh. S.K.Paul the then Agri. Engineer (Deptt.of Agri., Punjab) produced

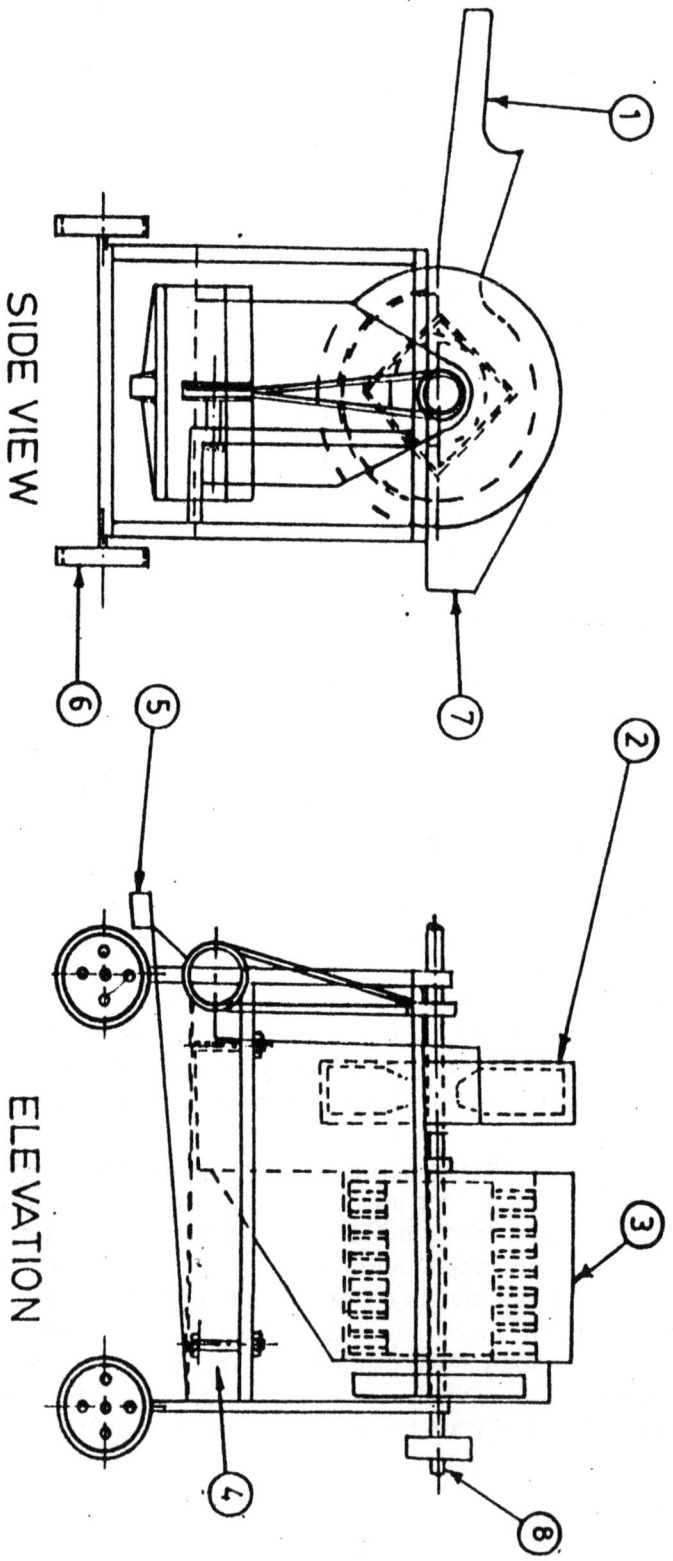

SIDE VIEW

ELEVATION

1 FEEDING CHUTE   2. BLOWER   3. THRESHING CYLINDER

4 OSCILLATING SIEVES   5. GRAIN OUTLET   6. TRANSPORT WHEELS

7. STRAW OUTLET   8. SHAFT

FIG.10.1 SCHEMATIC VIEW OF SPIKE-TOOTH TYPE THRESHER

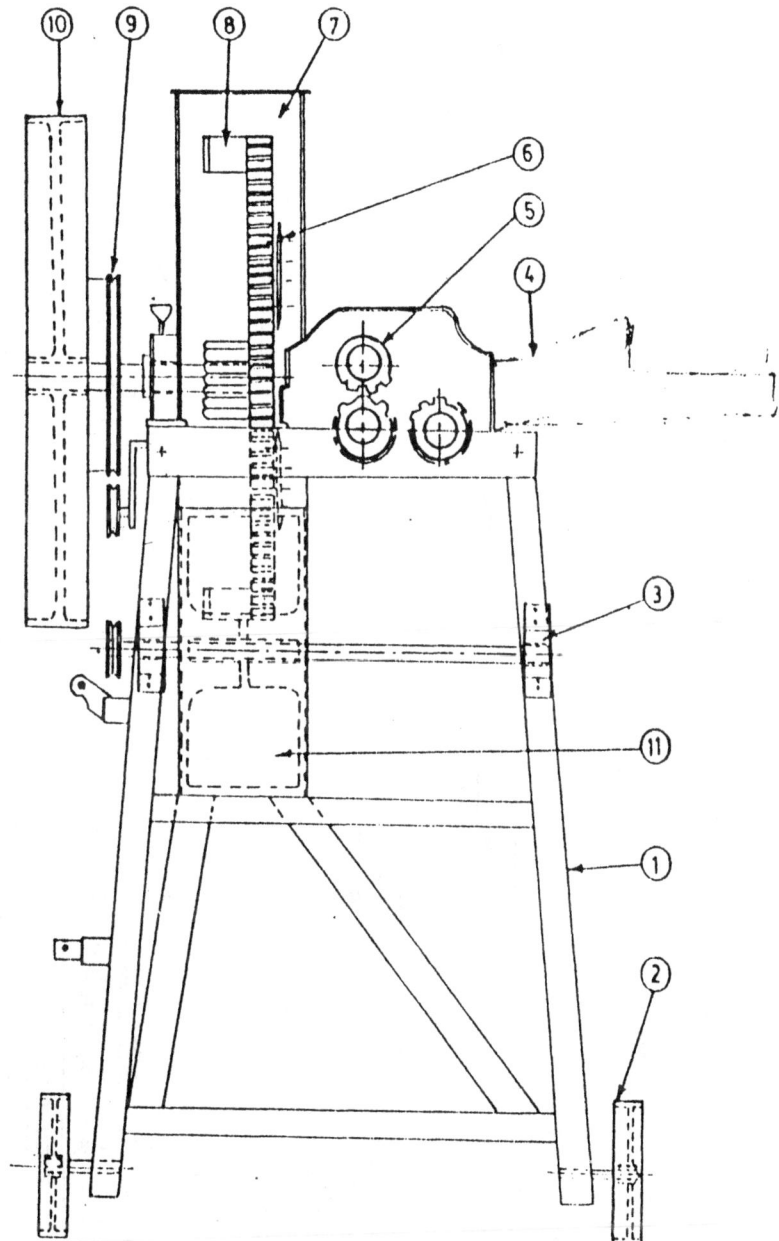

1. MAIN FRAME 2. TRANSPORTATION WHEEL 3. BEARING 4. FEEDING CHUTE
5. FEED REGULATING ROLLERS 6. CUTTING BLADE WITH TEETH 7. CYLINDER
8. BEATERS 9. V-BELT PULLEY 10. FLY WHEEL 11. FAN

FIG. 10.2 SCHEMATIC VIEW OF CHAFF-CUTTER TYPE
THRESHER

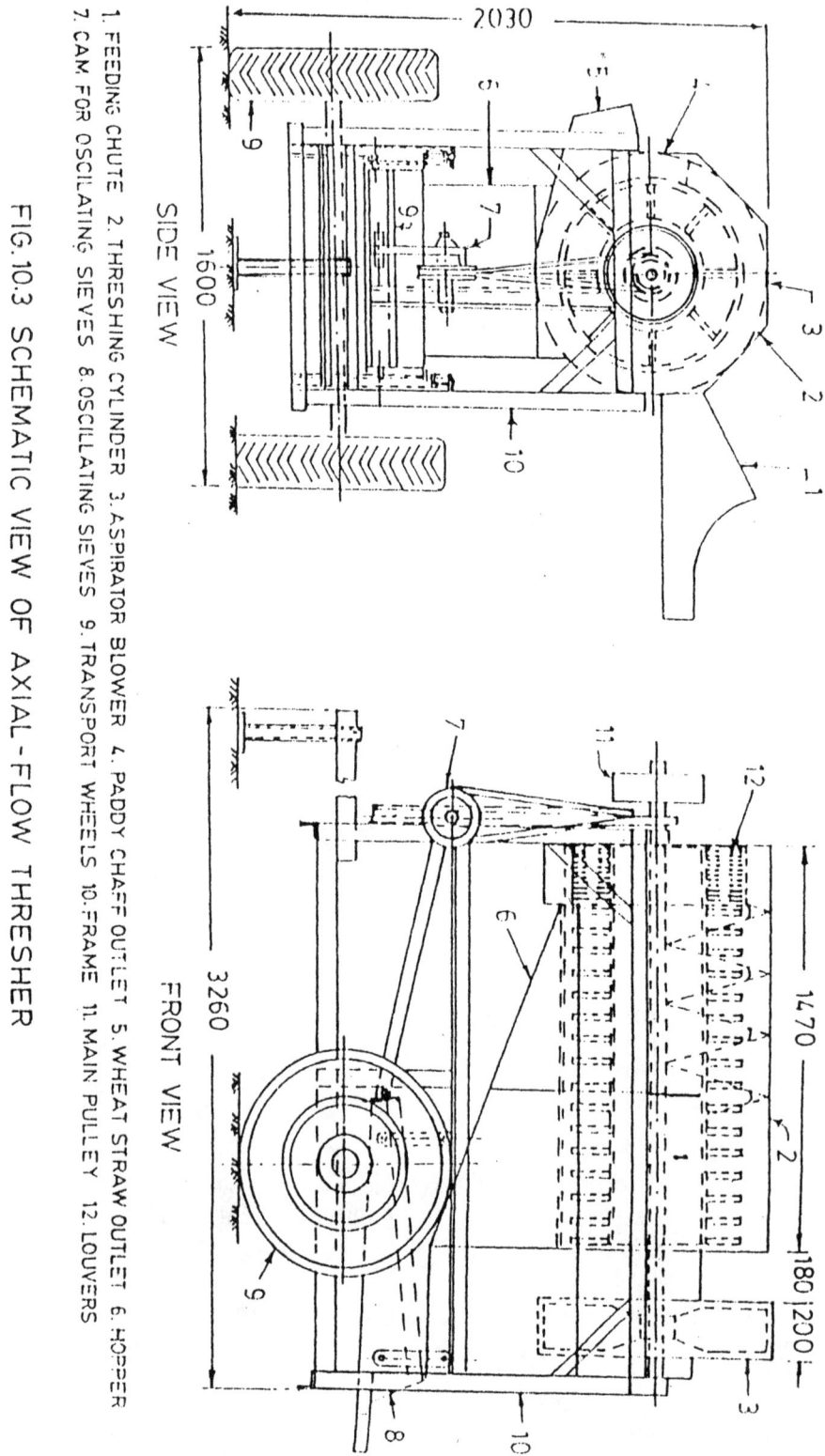

154

SIDE VIEW

2030

1600

FRONT VIEW

3260

1470

180 1200

FIG.10.3 SCHEMATIC VIEW OF AXIAL-FLOW THRESHER

1. FEEDING CHUTE 2. THRESHING CYLINDER 3. ASPIRATOR BLOWER 4. PADDY CHAFF OUTLET 5. WHEAT STRAW OUTLET 6. HOPPER
7. CAM FOR OSCILATING SIEVES 8. OSCILLATING SIEVES 9. TRANSPORT WHEELS 10. FRAME 11. MAIN PULLEY 12. LOUVERS

the first power wheat thresher in the mid fifties which could thresh, clean and bag the grains in a single operation. In this thresher, the threshing was carried out by hammer mill type threshing head and separation and cleaning was accomplished through the aspirating action of the air passing through the sieve by a blower which also sucked and flew away the 'bhusa' to some distance. Later on, the thresher with spike tooth type cylinder and a closed concave, aspirators and sieves was developed. Majority of the present day threshers have adopted the main design features of this thresher and have added accessories such as grain elevator for bagging, transport wheels, flywheel, automatic feeding etc.

There have been alarming cases of accidents during the use of power threshers. Critical analysis of the causes of accidents has revealed that about 73% are caused due to human factors such as carelessness, intoxication, etc., 13% due to machine factors like improper feeding system and use of inferior material of construction, 9% due to crop factors like abnormal crop and earhead threshing, and, 5% due to environmental factors like poor light during night operation, crowded surrounding or excessively hot weather conditions. With a view to control such accidents, the Govt. of India has enacted a legislation titled "Dangerous Machine (Regulation) Act-1983". According to this, every manufacturer of power thresher shall ensure that the machine and every part there of complies with the standard specified of BIS. Failure to comply with the conditions specified in the Act or Rules or Orders made there under, is liable to suspension/cancellation of the manufacturer's licence and other penalties as provided in the Act.

10.3 Principles of operation of thresher

1. Threshing

The threshing is accomplished by the impact of the rotating pegs or hammers (mounted on the cylinder) on the earheads which force out the grain from the sheath holding it. In the threshing of wheat crop, the straw is also bruised and broken up by the impact of the rotating tips, thus converting it into bhusa. Different grain crop and different varieties of the same grain crop have varying characteristics which require different speeds of the cylinder for achieving best results of threshing. Therefore, accurate adjustment of cylinder speed is essential. Some of the recommended speeds of cylinder and concave clearance for threshing different crops are given in Table-I.

2. Bhusa making

In the traditionally wheat growing areas of our country, wheat bhusa is an important roughage for cattle feed. Bhusa making is, therefore, an essential requirement of a thresher, particularly in the northern and central India and the major factor in deciding the selection of a make/model. The threshers developed in India for threshing of wheat are based on the principle of repeated beating to the crop in an enclosed

156

table 1   Recommended speeds of threshing cylinder for selected crops.

S No.	Crop	Cylinder speed (m/s)	Speed (rev/min)
1.	Wheat	20-30	550-1100
2	Paddy	15-25	675-1000
3	Jowar	12-20	400-675
4.	Bajra	10-16	400-550
5.	Gram	12-22	400-750
6	Peas	13-22	430-750
7.	Barley	20-26	740-1080

Source : Peter, 1986

threshing chamber till the straw is reduced to a size which would pass through the concave grate. In the syndicator thresher, the straw is chopped up on being fed axially in the threshing chamber by the chaffcutter blades which have serrated edge to prevent slippage of dry straw while shearing it. Generally, the length of cut is adjusted to about 20 mm. The cut straw is further crushed and bruised between the periphery of the corrugated flywheel and the concave grate or screen. The rubbing action completes the threshing operation.

The clearance between the rotating threshing cylinder and the concave has significant, effect on the threshing efficiency, the size of 'bhusa' obtained and the power consumption. However, a suitable combination of cylinder-concave clearance, concave grate/screen opening size and cylinder speed has to be arrived for different crops and their condition.

Generally, the concave is made of 6 mm square bars with grate openings of 6.35 mm, ranging upto 12.5 mm for various crops except maize which has recommended grate opening of 25 mm. In the peg type cylinder, the concave clearance is adjustable by changing the length of pegs. Table-2 gives the grate openings and concave clearance for threshing different crops.

Table 2 : Recommended range of concave clearance and grate opening for different crops.

S. No.	Crop	Concave clearance (mm)	Grate opening (mm)
1.	Wheat	15-25	6.35 - 8.5
2.	Paddy*	20-30	8.5 - 10
3.	Jowar	15-25	6.35 - 8.5
4	Bajra	15-25	6.35 - 8.5

5.	Gram	20-30	8.5 - 12.5
6.	Maize	20-30	25
7.	Barley	15-25	6.35 - 8.5
8.	Peas	20-30	8.5-10
9.	Soybean	20-30	8.5-10

----

* Paddy cannot be threshed with conventional wheat thresher used in India.
  Source : Peter, 1986

3. Cleaning

The threshed grain mixed with bhusa falling through the grate openings of the concave, has to be separated and cleaned. The aspirator system is most suited when the grains are mixed with "bhusa". The first stage of cleaning is done by the material falling through an upward, blowing column of air which carries away the lighter bhusa with it but allows the heavier grain and straw pieces to fall through. The air is blown by a blower whose inlet is connected to the aspirator column which carries the bhusa inside the casing of the blower and is blown out through the outlet. Air velocity of 4m/s to 5m/s is needed for the main aspirator to remove the straw which is already suspended in air. Only those pieces of straw which are large and nodes fall down on the sieve below.

The top shaker sieve is generally of 4.9 to 10.9 mm perforated sheet according to type of crop through which the grains pass down alongwith other smaller particles of soil etc. Larger pieces of straw and other materials like nodes slides over the top of the screen and are separated. The lower sieve is also a perforated sheet with hole size of 1.5 mm dia to 2 mm dia, through which the weed seeds of smaller size, soil and sand particles are fall out and only the grains alongwith some chaff remain on top, which slide out either for bagging or in most cases go under a secondary aspirator which sucks up the chaff and leaves the grains clean. The shaker sieves of the aspirator type cleaning system have a speed of 8m/min from a crank having a throw of 20 mm to 25 mm and rotating at a speed of 300 rpm to 400 rpm.

10.4 Terminology

a) Power thresher: A machine operated by a prime mover to separate the grains from straw.

b) Clean grain: Threshed, mature, unbroken grain and free from foreign matter.

c) Damaged grain: Threshed grain which is partially or wholly broken.

d) Unthreshed grain: Whole grain attached to straw after threshing.

e) Chaffed straw: Straw being discharged from threshing chamber which is

158

usually crushed cut.

f) Foreign material: Inorganic and organic material other than grain which includes sand, gravel, clay, metal chip, chaff and straw, weed and other inedible grains.

g) Grain Straw Ratio: Ratio of grain to straw by weight.

h) Maximum Input Capacity: The maximum feed rate at which no choking occurs in the thresher and no stalling occurs in the prime mover at the speed specified by the manufacturer.

i) Optimal Input Capacity: The feed rate at which efficiencies are within the specified limits of the relevant Indian Standards.

j) Output capacity: The mass of the grain mixture received at main grain outlet(s) when collected at optimal input capacity.

k) Percentage of blown grain: The clean grain lost alongwith chaffed straw 'bhusa' with respect to total grain input expressed as percentage by mass.

l) Percentage of broken grain: The broken grain from main grain outlet(s) with respect to total grain mixture received at main grain outlet(s) expressed as percentage by mass.

m) Percentage of spilled grain: The clean grain dropped through the sieve and overflown from sieve alongwith tailings with respect to total grain input, expressed as percentage by mass.

n) Percentage of unthreshed grain: The unthreshed grain from all outlets with respect to total grain input, expressed as percentage by mass.

o) Threshing efficiency: The threshed grain received from all outlets with respect to total grain input expressed as percentage by mass.

p) Transporting weight: The weight of thresher when transported.

q) Specified speed: The range of rotational speed of threshing drum-shaft when threshing wheat.

r) Specific thresher output: The mass of grain mixture per horse power per hour.

10.5 Type of test

The power threshers for testing are normally selected at random from the series production by a representative of the testing organization. The method of selection should be specified in the test report.

The following tests are conducted for thresher:

A. General Tests

i) Checking of specification
ii) Checking of material

B. Performance test at recommended speed

a) Input and output
b) Energy and power requirement
c) Losses through different outlets
d) Bhusa quality evaluation
e) Threshing efficiency
f) Cleaning efficiency
g) Performance at varying feed rates (optional)

C. Long run test

i) Effect on the performance at recommended speed over a long period of use.
ii) Convenience of handling, including transport
iii) General durability

10.6 Test procedure

Pre-test observations

i) Determination of grain-straw ratio:
Five samples. each weighing one kilogram of crop, will be taken and separated for grains and straw. Weigh the grains and straw separately for each sample and calculate their ratio. The average of five samples shall be taken as grain straw ratio.

ii) Moisture content of grain and straw:
Take suitable sample of grain and straw for moisture content analysis.

ii) Running-in and preliminary adjustments:
Before actual testing, the power thresher shall be run-in for atleast one hour. The adjustments for the speeds of different shafts, concave clearance, speed of prime mover, screen slope etc. shall be done as per manufacturer's recommendations.

crop description

Following characteristics of the crop used for thresher testing will be reported:

a) Name and variety of the crop
b) Moisture content of the grains
c) Moisture content of the straw
d) Length of plants (Avg. of 10 plants)
e) Grain-straw ratio

Verification of specifications

The main objective of this test is to study and measure preciously the specifications of the machine/critical components. The specifications not confirming to the values declared by the manufacturer as given in the relevant Indian Standard(s) need to be highlighted in the test report.

The following items are critically examined.

a) Design of threshing drum and its working parts such as peg tooth, bar, beater, etc.
b) Crop feeding mechanism
c) Mechanism of separation and cleaning
d) Provisions for adjustments, replacements and maintenance of working parts.
e) Balancing of threshing drum
f) Power transmission system
g) Safety provisions

10.7 Performance requirement

There is a wide spectrum of demands placed on the thresher, particularly the multicrop thresher. Grain moisture variation plays a significant role in determining the performance of threshers. The variation in the crop characteristics may call for precise adjustments. The main performance criteria for wheat threshing are given in
Table 3.

Table 3: Suggested acceptable limits of wheat threshers

S.No.	Item	Performance
1.	Threshing efficiency	> 99%
2.	Cleaning efficiency	> 96%
3.	Grain breakage	< 2%
4.	Total Grain loss	< 5%
5.	Split straw	> 95%
6.	Avg. length of Bhusa	12-15mm

Thresher output

The output of thresher depends on the condition of crop mainly in terms of crop moisture content and the grain-straw ratio. Generally a smaller thresher should give a better output due to the evenness of feeding. Also an electric motor operated thresher should give better results as there are less fluctuations in the speed of operation because of the better overload carrying capacity of the electric motor. Generally the hammer mill type of threshers have output of wheat grain varying from 15 kg to 20 kg per HP hour, the peg tooth type from 20 kg to 30 kg per HP hour and chaff cutter type threshers from 35 kg to 45 kg per HP hour. For reference purpose, the output for different crops of a typical thresher is given in Table 4.

Table 4 : Specific thresher output for different crops

S.No.	Crop	Output (kg/HP hour)
1.	Wheat	20-45
2.	Paddy	40-70
3.	Jowar	50-60
4.	Bajra	40-55
5.	Gram	30-40
6.	Maize	150-400
7.	Barley	25-50
8.	Peas	20-30
9.	Soybean	15-25

Source: Peter, 1986

The capacities of thresher depend on the power available from the prime mover and the rate of continuous feeding of the crop. The chute type inlet has lesser feeding capacity compared to the positive feed roller type, the hopper type and the feeding conveyor type.

Operation and collection of data at recommended speed

Install the thresher on level ground preferably on a hard surface and set the clearances, screen slope, etc. in accordance with manufacturer's recommendations. Attach the thresher with a suitable prime mover preferably an electric motor fitted with an energy meter. The data be recorded and analyzed as per Appendix-XVI and XVII.

Power consumption

Run the thresher at no-load for atleast 10 minutes at the specified recommended revolution of threshing unit and record the readings of the energy meter. Calculate the power consumption at no load for one hour. Similarly, calculate power consumption on load (mini. 3 readings) at maximum input capacity in order to find out energy/power requirements.

Collection of sample

During the test period at the maximum input capacity and at recommended speed, collect the following samples:

Three set of samples at an interval of 20 min. at the following outlets:

Main grain outlet	60 seconds
Sieve over flow	60 seconds
Bhusa outlet	10 seconds

The actual time for collection of samples should be recorded.

Analysis of sample

Main grain outlet : Analyze for clean, unthreshed, broken grains and foreign materials on the basis of 100 gm sample.

Bhusa outlet : Analyze for unthreshed and clean grains

Sieve overflow : Analyze for unthreshed and cleangrains

Sieve under flow : Take 1/60th of the quantity of the material dropped through the sieve and analyze for clean unthreshed grains.

Performance at varying speed

This test is performed to find out the effect of speed on the performance of the thresher. Generally, two speeds higher and two lower than the recommended are taken and performance observed on the same lines as in the preceding section.

Performance at varying feed rates

Effect of feed rate on the performance is observed by feeding the machine at different feed rates. Minimum three feed rates should be selected suitably based upon the capacity and tested.

Field tests

The thresher is run continuously for a total atleast 20 hours. The duration of each run should be about 3-5 hours. The thresher is run using an electric notes of recommended horsepower.

Samples from the bhusa outlet, grain outlet and tailings outlet, sieve overflow and sieve underflow are taken after every hour to check the performance. Energy consumption of the motor is also recorded. The convenience in making field adjustments, service, maintenance, change of parts and transport, etc. are carefully observed and reported highlighting the differences, if any.

Any breakdown and major adjustments required during the test (both laboratory and field) are reported. After the test, major components are opened to check for any abnormal wear and tear which must be highlighted in the report.

10.8 Computation of results

A. Thresher main pulley size

The size of the main pulley to be fitted on the thresher can be calculated using the following formula:

$$N_1 D_1 = N_2 D_2$$

$$D_1 = \frac{N_2 D_2}{N_1}$$

Where :

N_2 = Speed of thresher pulley rev/min
N_1 = Speed of prime mover pulley rev/min
D_1 = Diameter of thresher pulley (mm)
D_2 = Diameter of prime mover pulley (mm)

B. Performance parameters

From the analysis of samples and sampling time, the performance parameters are calculated as follows:

(i) Total grain input

$$A = B + C + D$$

A = Total grain input per unit time by weight
B = Weight of threshed grain (whole and damaged grain) per unit time collected at the main grain outlet
C = Weight of clean grain per unit time collected from sieve overflow, sieve under flow and bhusa outlet
D = Weight of unthreshed grain from all outlets per unit time

(ii) Percentage of damaged/broken grain = $\dfrac{E}{F} \times 100\%$

E = Quantity of damaged/broken grain in the sample taken at main grain outlet

F = Total quantity of sample taken at main grain outlet

(iii) Percentage of blown grain = $\dfrac{G}{A}$ x 100%

G : Quantity of whole grain collected at bhusa outlet per unit time

(iv) Percentage of spilled grain = $\dfrac{K}{A}$ x 100%

K = Quantity of clean grain obtained at sieve overflow and underflow per unit time

(v) Percentage of unthreshed grain = $\dfrac{H}{A}$ x 100%

H : Weight of unthreshed grain per unit time obtained from all outlets

c) Efficiencies

i) Threshing efficiency(%) = 100-percentage of unthreshed grain

ii) Cleaning efficiency(%) = $\dfrac{I}{F}$ x 100

I: Weight of whole grain per unit time at the main grain outlet

F: Weight of whole material per unit time at the main outlet

10.8 Determination of corrected output capacity

To avoid the variation of moisture content of grain, the output capacity obtained may be corrected when the standard moisture content and grain ratio can be specified and usually taken as 12 percent and 40 percent respectively and calculated by following formula:

$$Wc = \left(W - \dfrac{W(M-12)}{88}\right) \dfrac{40}{R}$$

Where :

Wc = Corrected output capacity
W = Output capacity obtained
M = Observed moisture content of grain
R = Observed grain ratio in percent

10.9 Determination of power consumption

i) In case of energy meter fitted prime mover, the difference between no load and on load readings shall give power consumption. Calculate the correct power consumption for one hour giving due allowance for drive losses as 6 and 3 percent for flat belt and V-belt drive respectively.

ii) In case of dynamometer fitted prime mover, the average of reading taken shall give the average torque required. Calculate power requirement by following formula:

$$\text{Power (Kw)} = \frac{\text{Torque (kgf-m) x speed (rev/min)}}{973.363}$$

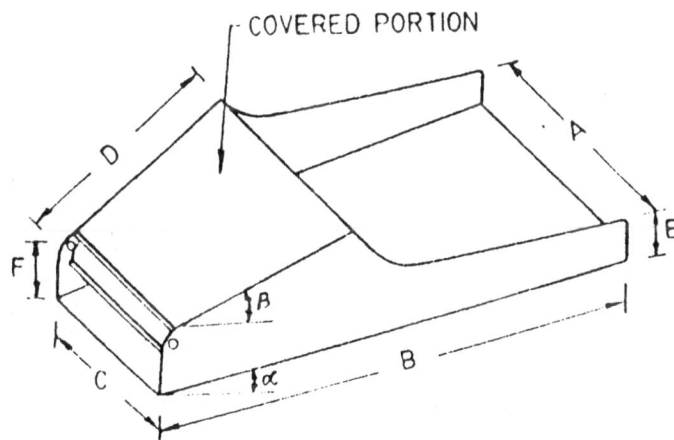

FIG 10.4 SAFE FEEDING CHUTE

10.11 Checking for safety

Feeding chute provided should be checked from safety point of view. The dimensions of chute be checked as per diagram shown in Fig. 10.4 and compared with the relevant standards. However the recommended range of various dimensions for different type of threshers is given below:

A	=	450 to 650 mm
B	=	Mini. 900 mm
C	=	350 to 500 mm
D	=	Min. 450 mm
E	=	50 to 60 mm
F	=	125 to 200 mm
/α	=	10" to 15"
/β	=	10' to 30"

Chapter 1I

DATA ACQUISITION, PROCESSING AND ANALYSIS

11.1 Introduction

There have been significant developments during the last 2 decades in the field of instrumentation. Recently, use of personnel Computers in testing and evaluation of agricultural machinery has revolutionized the old concept and consequently has brought about tremendous saving in time and labour together with the highest degree of accuracy. Therefore, computer based instrumentation has great potential in development and standardization of agricultural. machinery. There are three basic functional elements which are an integral part of instrumentation system namely data acquisition, measuring devices and data analysis.

11.2. Data acquisition

Data acquisition systems are required for recording of large scale data. For instance the input signals like temperature, pressure, speed, flow rate etc. may have to be recorded periodically or continuously. In such a situation, a data acquisition system can be used in which the digital output of a number of input signals can be recorded. Fig 11.1 shows that an input scanner scans the input signals one after the other, which after amplification are converted to digital form by A-D converter. The digital computer or microprocessor based micro computers depending on the number of channels can be used which are commercially available for laboratory as well as industrial purposes.

Basically we have to see the characteristics of the signals. Signal is a physical quantity, which has something special as it transfers data which can be observed and is usually changed to electric signals by appropriate transducers.

A- Type of signals

i) Random signals

    a- Sound or voice
    b- Earth quake
    c- Atmospheric temperature
    d- Coarseness of metal surface

Fig. 11.2 indicates that the variable which is shown on the horizontal axis of (a),(b), and (c) is the time and another variable shown on the horizontal axis of (d) is position of metal surface i.e. distance. Although the value of these signals are confirmed on specific time but they can not be fixed afterwards. Therefore, these signals are called random signals.

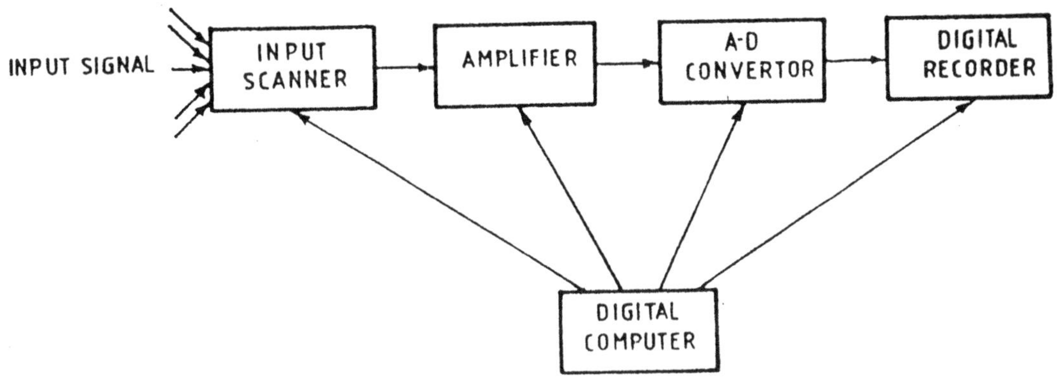

FIG.11.1 BASIC ELEMENTS OF DATA ACQUISITION SYSTEM

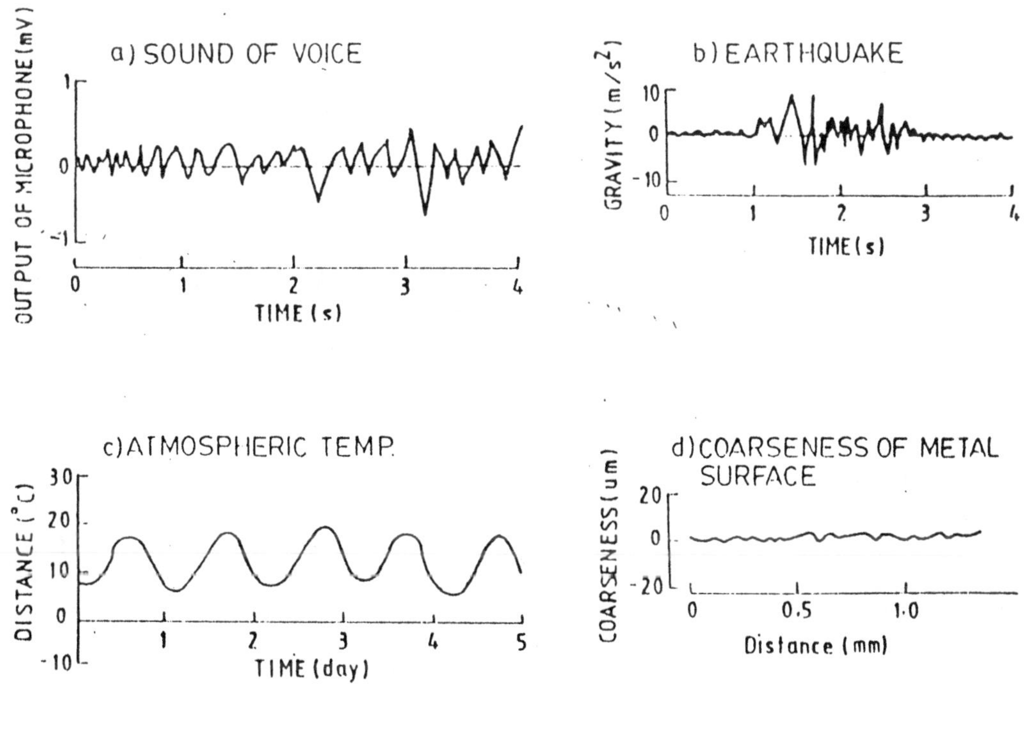

FIG.11.2 SHAPE OF DIFFERENT SIGNALS

ii) Deterministic signals

Those signals whose value can be fixed definitely at any instant of time are call deterministic signals e.g. sine wave.

Function of sine wave f(t) is defined as :
F (t)   = A sin (wt+θ)

Where:

$$A = \text{Amplitude of signal}$$
$$W = \text{Angular frequency}$$
$$t = \text{time}$$
$$\theta = \text{Phase angle}$$

Drawing the same wave at regular intervals, signals such as sine wave as shown in Fig. 11.3 (a) have the name of periodic signal and is represented by the following equation

f(t+ nT) + f (t)

Where

$$T = \text{time period} + \frac{2}{w}$$

n   = 1,2..................

iii) Pulse signal

Those signals which are concentrated at very short time are called pulse signal as shown in Fig 11.3 (b) e.g. wave energy. In such signals, pulse are counted per unit time.

B) Conversion of signal (analog to digital)

Analog instruments are those which present physical variables of interest in the form of continuous or stepless variations with respect to time. Whereas digital instruments are those in which the physical variables are represented by digital quantities. The value of signals altering continuously are given the name of "analog signal". The action of converting analog signals to the signals with interval is called "dispersion".

The device which converts the analog value to digital value is called A/D (Analog to Digital)converter. In many A/D converters, input is usually voltage, and output is the result of A/D conversion which can be managed by the computer. Even if high performance A/D converter is used, some noise comes into play, in the connecting parts due to electric potential difference between the ground of signal source and the

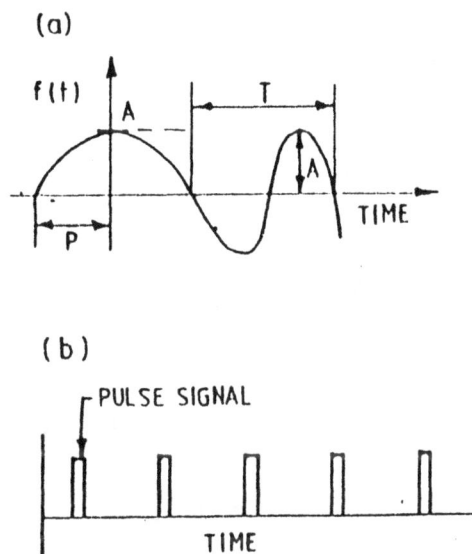

(a)

(b)

PULSE SIGNAL

FIG.11.3 SHAPE OF SINE WAVE (a) AND
PULSE WAVE (b)

ground of A/D converter. shielded wires are usually used to connect analog signal source with A/D converter. In order to neutralize the electric potential difference. double or triple lined shielded wires should be used for connecting analog signal source with A/D converter.

Normally, the analog to digital conversion is carried out in two steps. In the first step, the analog data is separated by sampling the data after a definite time interval (t_s) known as sampling time. In the second step, the corresponding digital value is assigned a 4-bit binary code so that conversion becomes compatible with the codes used in the digital computers as shown in Fig. 11.4.

C) Sampling of signal

Sampling is necessary for the data to be analyzed, usually at equally spaced time intervals so that discrete samples are produced and fed to the digital computer. The samples are taken at a rate that is at least twice the highest frequency present in the wave form. When converting sine wave to digital value, it is necessary to use such sampling as shorter than half of the period of the wave, in order to reproduce precisely original wave from converted effects. The frequency two times bigger

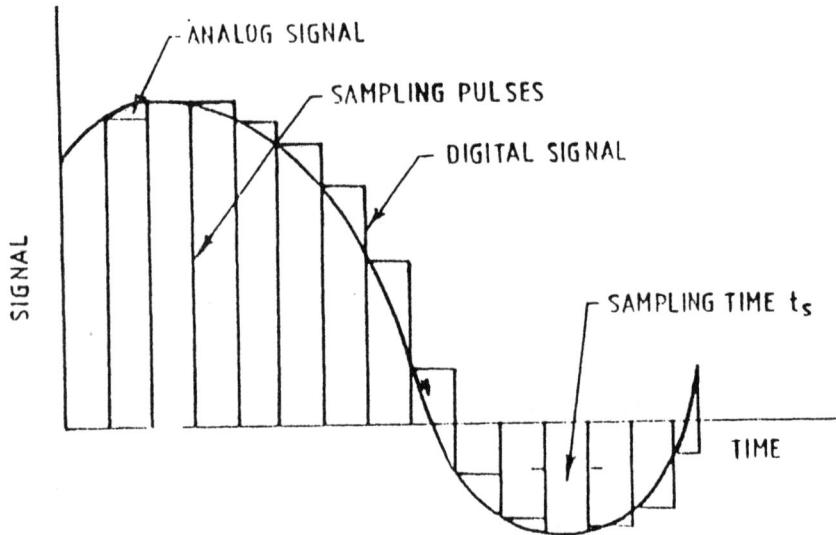

FIG.11.4 TYPICAL ANALOG SIGNAL SAMPLING FOR CORRESPONDING DIGITAL VALUE

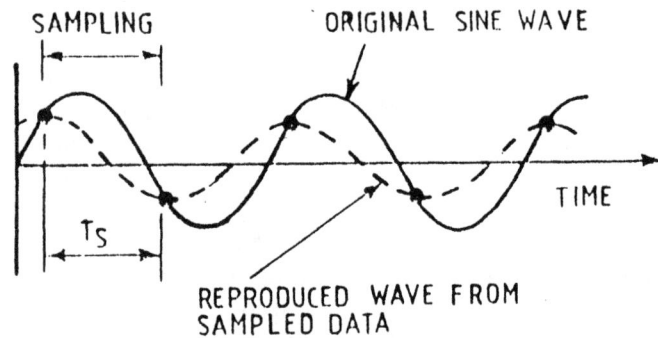

FIG.11.5 SAMPLING OF CONTINUOUS DATA AND ALIASING

than the frequency of the original wave is called "nyquist frequency". Figure-11.5 shows sampled sine wave under the sampling frequency of less than Nyquist frequency. In this figure, we can notice the lower frequency which has not been included in the original wave. This phenomenon is known as aliasing. However two sine waves if sampled at interval (Ts) would give the same impression as far as the nature of wave is concerned even though one sine wave has higher frequency than the other.

Low Pass filters with amplifier can be used for carrying out digital operations on various samples. These will remove even unwanted noise and vibration.

11.3 Measuring devices

A-   Types and kinds of sensors
   i) Load cell
   ii) Magnetic pulse pickup
   iii)Photo sensor
   iv)Rotary encoder
   v) Thermo couple etc.

B-   Types and kinds of physical devices
   i) Pulse counter:
        a- High range pulse counter
        b- Low range pulse counter

   ii) Strain amplifier
   iii) Smoke meter
   v) Torque meter
   v) Fuel flow meter
   vi) F/V converter (frequency to voltage converter)
   vii) A/D converter(analogue to digital converter)
   viii) Sound level meter
   ix) Vibration meter etc.

11.4 Data analysis

Modern electronic digital personal computers are playing a vital role in the advancement of science and technology. This is due to their ability of processing a large amount of information and at fantastic speed, precision and accuracy. Presently, computers are being conveniently used in a vast variety of applications like in the design and analysis of tractors and agricultural machinery, cost effective study of complicated systems, project planning and decision making, behavioural and environmental analysis, business accounting and inventory control, personnel management and data processing etc. Further it can handle a large number of significant digits in its arithmetic operation and thus can provide extremely accurate results. In addition, the computer has a perfect memory and also has the capability to automatically retrieve the stored information and independently execute the instructions.

It may be interesting to note that a computer can neither think nor judge on its own. It is like an obedient servant waiting to be told what to do. Therefore, in order to perform a specific computational task, the computer has to be precisely instructed when to start, what data to use, what steps to follow and when to stop.

11.5 General approach in computer problem solving

Techniques to achieve the objective quickly using optimally the available resources are the most important part of the whole operation of utilizing the computer for data analysis or any other problem. Therefore, it is imperative to adopt a systematic, logical and step-by step approach for solving any problem by means of a computer as listed below.

1. Problem identification
2. Mathematical formulation
3. Problem planning and flow charting
4. Algorithmic formulation (i.e. precise and unambiguous statement of actions to be carried out by the computer)
5. Programme writing
6. Programme check out
7. Printing of the results

11.6 Computer components

The IBM (or compatible)Personal Computer(also called PC) is made up of two things:

a. Hardware- the physical equipment
b. Software- the programmes that tell the computer what to do

a) Hardware

Every PC has hardware items like the computer itself, the keyboard, a video display or television set, cassette tape recorder or a disk drive.

The cassette recorder and the disc drive do the same function. But the disc drive is much more convenient to use. Optional hardware items include printer, modem (communicating by telephone) etc.

i) Computer

The computer itself is a rectangular gray and white typewriter-sized object as shown in Fig. 11.6. It has a flat top, air vent in front and back, and a recess that holds the disc drives. Although we are not concerned about the inside of computer. But for sake of information, the computer has the following items.

a) A brain, usually called a central processing unit(CPU). Sometimes it is referred as microprocessor.

b) Memory for remembering data and programmes. Computer memory is defined in units called bytes. Each byte can hold one typed character. PCS have many

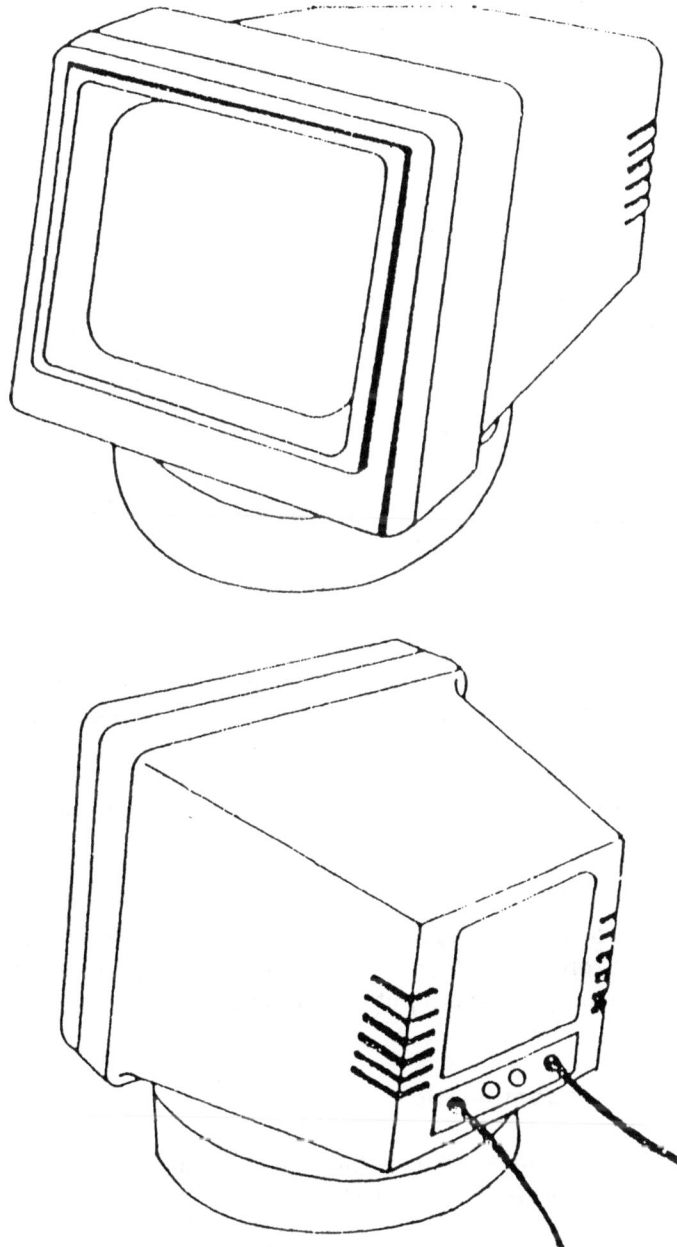

FIG.11.6 SCHEMATIC VIEW OF PERSONNEL
COMPUTER

174

thousands of bytes of memory, so we refer to the total memory in terms of Ks (1 K = 1024 bytes). PCs usually come with 640 K.

c) Connections for the attached devices, such as video display, keyboard and cassette recorder etc. These connections act like the body's muscles and nerves and known as interfaces.

ii) Keyboard

The PC does not have a built-in keyboard. Instead, the keyboard is separate and connected to the computer through a coiled cable. This separation helps to hold the keyboard in lap or move it around. We can also unplug the keyboard to prevent form children and other curiosity seekers from using it. The keys are arranged in three groups, as shown below :

1- The function keys
2- Typewriter like key
3- The numeric/cursor control key pad

A typical keyboard is shown in Fig. 11.7.

iii) Display

There are several different kinds of video displays (like television sets)that may be attached to PC. We may have an monochrome display, colour display monitor or a standard television set or television monitor. For typing purpose all video displays are same. They show 25 lines of 80 characters each (25 rows and 80 columns). In reality, only 24 lines are available as the computer uses the bottom line for its own special purposes.

To draw pictures of figures (graphics) with computer, the various displays have different characteristics. The colour/graphics monitor or television set controls tiny dots (320 or 640 horizontally, 200 or 480 vertically) to create highly detailed pictures. The monochrome display allows to put simple letter-sized shapes in the normal 25-by-80 character grid.

Keyboard and display combination

The keyboard and display operate as a unit. This combination, in fact, is called video terminal. The special features it has are:-

a- The video display does not provide a permanent record of what it shows. Of course, this is blessing, since it saves paper and keeps the operation nice and clean. However, in book keeping or accounting, we need permanent record. Besides, it is awkward to carry the display around for consultation or mailing to some one. Therefore, printer is the solution of problem.

175

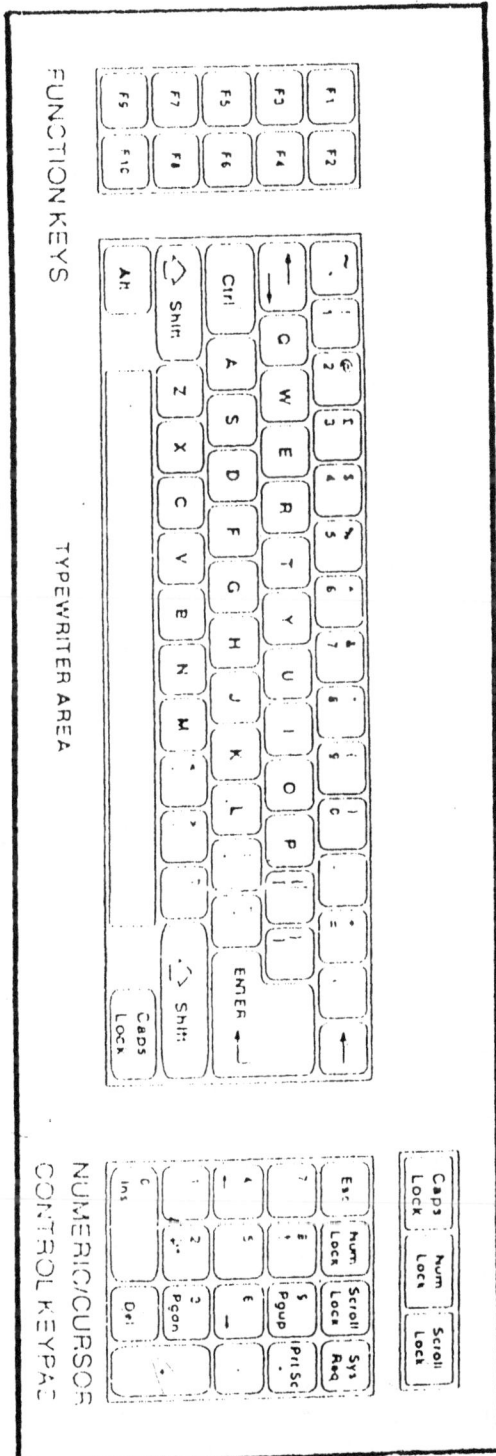

FIG. 11.7 A TYPICAL KEYBOARD

b- The video display has a limited areas. It allows to see only a certain amount (sometimes called a window) of material at any particular time. After it prints out the bottom line, it automatically moves the entire screen up by one line to make room for another line. Then the top line is lost from view. This movement is called scrolling.

c- A video display has characteristics different from those of a typewriter. The erasing character is much simpler, since no permanent record has been made. Similarly the entire lines or sections can be moved up or down easily and much faster on a screen than on paper. Therefore the video display can do everything that a television can do and at the same speed.

d- The new characteristics and freedom of the video display require additional control keys and markers. On a typewriter we can see where the carriage or typing mechanism is. Whereas computers solve this problem by providing a special marker or cursor that indicates the current position. We can rapidly move the cursor anywhere on the screen up or down, all the way to the right or left. The cursor can be moved by pressing the key available on the numeric keypad.

iv) Mass Storage Devices

Human being have permanent memory. Once they fully learn how to read and write or learn, they are embedded in their brain and not lost. Whereas a computer develops amnesia at the moment we switch off the power. Thus its memory is erased completely, and it forgets everything it ever knew. Hence by switching off the power, the current work is lost. Therefore, computer needs a way to store data and programme so that informations can be loaded into the computer's memory. For this mass storage devices are used. The PC can emply two types of mass storage devices:

i) Cassette recorders
ii) Disk drives.

Cassette recorders are slow and inconvenient to use. Therefore most PC owners prefer disc storage. Therefore, generally atleast one disk drive is made available with each personnel computer.

v) Disk Drives

The PC can house two disk drive. Disk drive acts much like record players accept that they play thin. Flexible disks are called floppy disks (or sometimes Just disks). The PC disks are 5.25 inches or 3.5 inches in diameter. The PC refers to its disk drives as A and B. Generally, A is the main drive, and B is the secondary drive. Like tape recorders, disk drives record information on a magnetic surface. But to read something from a tape, a recorder must wind or rewind the tape until it reaches the correct position. Whereas winding or rewinding is not required in disk drives. Therefore, by disk drives we can read much faster than with tape.

vi) Printer

To produce a written record of results or programmes, we must have a printer. The two most common types of printers are:-

i)   Dot matrix printer
ii) Laser printer

a-   Dot matrix printers: They form characters by making patterns of dots in a rectangular grid. Dot matrix printers are generally fast and inexpensive, because of the gaps between the dots. The more dots in the printer's grid, the better its output looks, since the gaps are smaller. Dot matrix printers also have the advantage that they can produce plots and pictures as they can create any pattern of dots.

b-   Laser printers: A laser printer works like a copying machine. A laser printer can actually imprint entire characters on the paper, draw pictures and can obtain different type of faces and character sets (e.g. Greek letters or mathematical symbols). In general, laser printers are more expensive than dot matrix printers.

viii) Modems

Modems are devices through which one computer can talk to another over a telephone line. A modem is necessary to use electronic bulletin boards or information services. Electronic bulletin boards allow subscribers to advertise goods and services to other subscribers via computer. Such information services can provide stock market quotations, the front pages of major newspapers, sports scores, restaurant and hotel accommodations etc.

11.7 Software

Generally a computer manufacturer provides few programms with the PC which are built in the computer. These programms are:-

i)   Basic input/output system (BIOS)

The internal programme that controls the PC's overall operation are the basic input/ output system or BIOS. Input is simply fed into the computer and output comes out of it. The BIOS interprets words typed from the keyboard, displays characters on the screen, and transfer data to and from the disk drives, printer, and whatever else is attached to the computer. In nut shell without BIOS, the PC would be just a show piece. BIOS is the PC's native intelligence that makes it a computer.

ii) Diagnostics

Diagnostics programme tests the computer, keyboard, display and printer to help in solving operating problems. The idea here is to reduce to take computer to dealer for any repair etc.

iii) Word Processors

For scientist, Engineers, writers or anyone else who has to prepare reports. there are word processors and spelling checkers. A word processor permits the user to display text on the screen, correct it, change it and even move words and paragraphs. Popular word processors include word star, word perfect and essay writer. These programms are handy for producing form letters, creating legal documents for standard paragraphs and developing new proposals or reports from previous work.

iv) Entertainment Programms

Although PC is not primarily intended for playing games, however many games are available for it. These include space games, war games and adventure games etc. etc.

11.8 Operating systems

If PC has a disc drive then we shall need disk operating system(DOS). DOS is a collection of programmes that makes it possible to move information to and from the disk. Thus, DOS acts as a computer's chief administrator. The DOS disk also includes BASIC intended specifically for disk systems.

Other computer languages

There may be certain applications for which BASIC is not well suited. So some other computer languages which are alternatives to BASIC include C. FORTRAN. PASCAL. and assembly language etc.

11.9 Writing of programmes

Sometimes we may not have ready programme that solves out problem. If the problem is large, then probably we may have to wait for a programme that can be bought. However, if the problem is simple, such as calculating of power or specific fuel consumption from measured data than we may have to write our own programme. Developing a computer programme involves the same thought processes as preparing a tax return or writing a book report. Thus we must decide that what to do, get the right tools and do it. In fact. most computer user buy more programmes than they write. Writing a programms is not terribly difficult, but requires skill and is time consuming. Anyhow, we can use a computer productively without writing any programme, as we can be an actor without writing scripts.

Chapter 12

SAFETY TESTING OF AGRICULTURAL MACHINERY.

12.1 Introduction

Tractors and agricultural machines need to be designed and constructed in such a manner that they do not danger the safety of the operator when properly used. However to achieve this. every operation and maintenance of the machine needs to be carried out in accordance with the manufacturer's instructions. The basic safety requirements ought to be met by proper design of the machine components. Additionally the machine needs to be equipped with special provisions. These may include for ensuring safety on guards, shield, safe location of the dangerous parts, safety cabins and frames etc. Functional components require to be kept anti-roll exposed for proper operation. need to be shielded to the maximum extent permitted by the intended function of the components. Additionally, under certain circumstances. warning signals and labels against possible danger/hazard need to be provided on the machine. Therefore. testing and evaluation of safety provisions on tractors and agricultural machinery assumes special significance in order to keep down the accidents to the minimum.

12.2 Types and causes of agricultural machinery accidents

Overturns : Machine working at steep slopes. high speed and quick turning, quick starting etc.

Run over : Person inspecting without engine stopping

Trap: Person wearing loose clothes wrapped in moving belts. chains. shafts etc.

Cut : Body parts touching the cutter knives. sharp edges etc.

Crush : Person seating/standing between machines and attachments; machines and walls. under attachments etc.

Fall : Person slipping from platforms. step.s

Burn : Body part touching on the exhaust pipes.

Fire : Careless use of fire, smoking. exhaust, carbon particle etc.

Health hazard : Noise, vibration, pesticide spray etc.

Hits : Thrown objects .

12.3 Technical requirements for ensuring the machinery safety

a) Safety guards

It is a protective device designed and fitted to minimize the possibility with machinery hazards as well as to restrict access to other hazardous areas. Different types of guards are shown in Fig. 12.1.

b) Safe distance

A means of providing guard where the possibility of contact with the hazard is minimized by the combination of the guard configuration including openings. The separation dimensions of pinch points in relation to body parts should be as shown in Fig. 12.2 and 12.3.

c) Safety devices

A device provided to minimize the incidence of machinery hazards like unexpected movement e.g. safety starting device, emergency stopping switch etc.

d) Safety signs

Information affixed on the machine to alert persons to hazards which can cause personal injury. The sign/caution plate should be permanently fitted/attached to the machine so that it can be readily seen and should not be easily removed.

e) Operational ease

The operating controls shall be so designed and installed that they can be operated safely and easily from the operator's seat. Maximum actuating force required to operate various controls are shown in Table-I.

12.4 Testing of agricultural machinery for safety

The main objective of safety testing is to ensure the safety of machinery operator and prevent accidents. All technical safety requirements as specified by the standards be checked thoroughly and reported.

12.5 Checking tools and devices

i) Scales:Checking of machine dimensions, safety distance and opening etc.

ii) Push pull tester: Checking of actuating forces required to operate contols.

iii) Sound level meters: Checking of noise level at operator's ear level.

iv) Vibration level meter:Checking of vibration level of operating controls and workplaces.

v) Surface thermometer:Checking of surface temperature of hot parts.

Table 1 : Maximum Actuating Force Required to Operate Control as per ISO
Recommendation

Control	Type of control	Maximum actuating force to operate control(N)	Type	Remarks
Service brake	Pedal	600	Pressure	It should be
	Hand lever	400	Traction	possible to
				achieve effec-
Parking brake	Pedal	600	Pressure	tive braking
	Hand lever	400	Traction	performance
				when these
				forces are
				applied.
Clutch		350	Pressure	
Dual clutch	Pedal	400	Pressure	
Power take-off	Pedal	300	Pressure	
Coupling	Hand lever	200	Traction	
Manual steering system	Steering wheel	250		Force required to achieve a
Power-assisted steering system with failure of the power-assisted steering force.	-do-	600		turning circle of 12m radius
Hydraulic power lift system	Hand lever	70	Pressure & Traction	

12.6 Methods of safety testing

Parts/devices and systems required

a) Guards for moving parts

i) Parts to be guarded in order to prevent danger to persons

- All shafts (including joints, shaft ends and crank shafts, universal joints, keys, pins and set screws etc. that protrude from moving parts.

- Pulleys, flywheels, gears, cables, chains, sprockets, belts, clutches and couplings

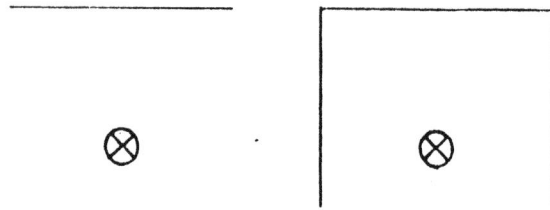

1. SHIELD AND COVER : PROTECTION OF THE SIDE
OR SIDES

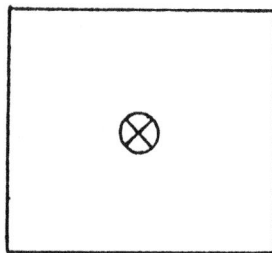

2. CASING : PROTECTION OF
ALL SIDES

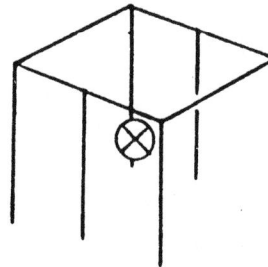

3. ENCLOSURE : PROTECTION BY RAILS,
FENCES, FRAMES ETC.

FIG. 12.1 DIFFERENT TYPES OF GUARDS

- Working parts like rotary tynes, digging blades cutting, binding, cutter knives and conveyors etc.

- Ground wheels, tyres and track adjacent to the operator's position.

ii) For guards formed out of a mesh or grill, the permissible size of openings are as follows:

I) For rectangular opening
$D \geq 850$ and $x \leq 135$ where D = distance between the guard moving
$D \geq 200$ and $x \leq 30$ part(mm)
$D \geq 120$ and $x \leq 20$ X = width or dia of opening(mm)
$D \geq 15$ and $x \leq 8$
$D \leq 15$ and $x \leq 6$

1) RECTANGLE OR SLOT

LIMB	ILLUSTRA-TION	WIDTH OF APERTURE a	SAFETY DISTANCE TO DANGER SOURCE b
FINGER TIP		$4 < a \leqslant 8$	$b \geqslant 15$
FINGER		$8 < a \leqslant 20$	$b \geqslant 120$
HAND	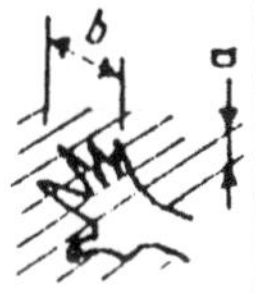	$20 < a \leqslant 30$	$b \geqslant 200$
ARM	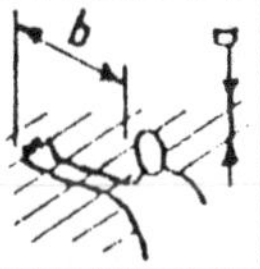	$30 < a \leqslant 135$	$b \geqslant 850$

2) SQUARE OR CIRCLE

LIMB	ILLUSTRA-TION	WIDTH OF APERTURE a	SAFETY DISTANCE TO DANGER SOURCE b
FINGR TIP		$4 < a \leqslant 8$	$b \geqslant 15$
FINGER	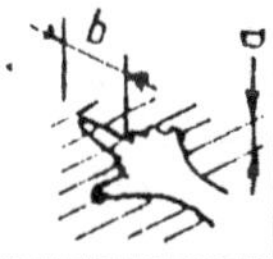	$8 < a \leqslant 25$	$b \geqslant 120$
HAND	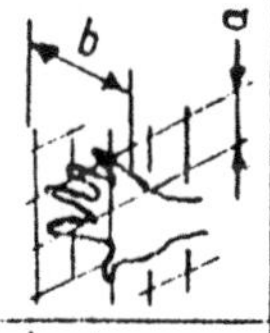	$25 < a \leqslant 40$	$b \geqslant 200$
ARM		$40 < a \leqslant 250$	$b \geqslant 850$

(Unit: mm)

FIG.2. SAFE REACH DIMENSIONS THROUGH OPENINGS

LIMB	BODY	LEG	FOOT	ARM	HAND WRIST FIST	FINGER
ILLUSTRATION			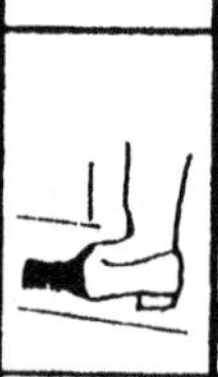			
MINIMUM SEPARATION DISTANCE REQUIRED	500	180	120	120	100	25

(ISO Unit : mm)

FIG. 12.3 MINIMUM SEPARATION DISTANCES FOR PINCHING POINTS

II) For circular or square opening

    $D \geq 850$ and $X \leq 250$
    $D \geq 200$ and $X \leq 40$
    $D \geq 120$ and $X \leq 25$
    $D \leq 15$ and $X \leq 8$
    $D \geq 15$ and $X \leq 6$

III) For Blower of air blast sprayer or liquid chemicals sprayer

    $D \geq 200$ and $x \leq 40$
    $D \geq 40$ and $x \leq D/5$
    $D \leq 40$ and $x \leq 8$

iii) Guards should be sufficiently strong.

b) Guards for PTO shafts

1. The PTO shaft should be guarded by casing cap, which should be firmly screwed or bolted to the machine body.

2. The PTO drive shafts as well as the universal shaft should be guarded by a casing throughout their length. The casing should be secured firmly and held in stationary position.

3 Guards should be sufficiently strong.

c)   Safety devices

1.Every stationary machine should have provision to disengage the power drive shaft. The control of the device should be located within easy reach of the operator.

2. A brush cutter stubble shave should be provided with adequate means of disengaging the power to the knife blades easily and promptly.

3. Combines harvestor should be equipped with a device to disengage the power of knife bar automatically in the event of clogging or fastening.

4. A portable type power unit such as the power knapsack sprayer should be provided with a quick release clutch to enable the operator to disengage the power.

5. All machine with lifting members should be provided with a locking device to keep the member in raised position.

6 Power driven machines should be equipped with a provision to stop the prime mover instantly

d) Braking device

1. Self propelled machines should be provided with both the service brake (main brake) and the parking brake.

2 The towed machine/equipment especially trolleys should be provided with the parking brake.

e) Operator's workplace

1. Any machine on which the presence of a worker or operator is necessary, should be provided with handles or handholds to ensure the safety and convenience of operator's mounting and dismounting.

2. Any machine on which the operator is required to sit should be provided with a comfortable seat and adequate footrest. Seat will adequately support the operator in all working and operating modes and prevent the operator from slipping off the seat.

3. Any platform on which the operator is required to stand during the operation of the machine should be level and have a non-slip surface. It should also be provided with guardrails around the platform.

f) Operating controls

The operating controls, such as steering wheel or lever, gear, selection lever, brake, clutch and switch should be arranged and fitted in such a way as to allow safe and easy control by an operator while standing or sitting in the normal operating position. The function and the operating method of these controls should be marked clearly.

g) Roll over protective structures

Safety cabs and safety frames (Roll over protective structures (ROPS) are now a mandatory requirement on all agricultural wheel tractors in all developed countries for the protection of the operator in the event of accidental overturning. The testing procedure has been described later in this chapter.

12.7 Definition of terms

i) Tractor mass

The mass of the tractor with full fuel tank as well as recommended coolant and lubricating oil, and with all components required for normal operation plus ROPS, but excluding optional ballast weights and operator.

ii) Tractor reference mass

The mass not less than the tractor defined in (i) above as determined by the manufacturer for calculation of the energy inputs.

iii) Tractor reference wheel base

A wheel base, not less than the maximum wheel base as determined by the manufacturer for calculation of the energy inputs.

iv) Operator seat

Any part of the seat or its structure including suspension and adjustment systems.

v) Seat reference points

It is that point where vertical line tangent to the most forward point at the longitudinal seat centerline of the seat back and a horizontal line tangent to the highest point of the seat cushion intersect in the longitudinal seat centerline section. It is determine with the seat unloaded and adjusted to the highest and most rearward position, as shown in Fig. 12.4.

vi) Vertical reference plane

It is the vertical plane, longitudinal to the tractor and passing through the seat reference point and the center of the steering wheel. Normally the vertical reference plane coincides with the median plane of the tractor.

vii) Zone of clearance

The roll over protective structure (ROPS) which may include overhead protection. fenders. cab sheet metal. or related ROPS parts outside. but near the operator area may be deformed in tests but shall not leave sharp edges exposed to the operator or

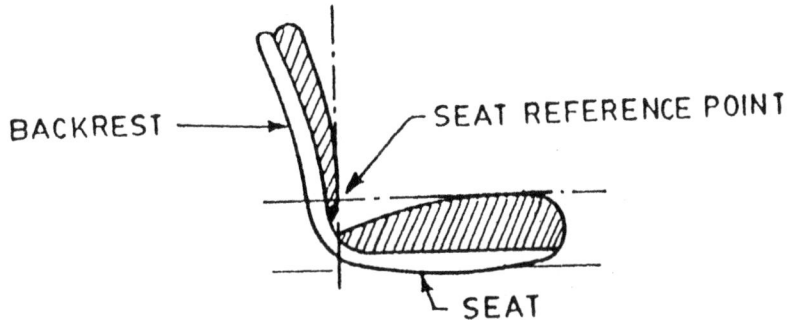

FIG.12.4 METHOD OF DETERMINING SEAT REFERENCE POINT

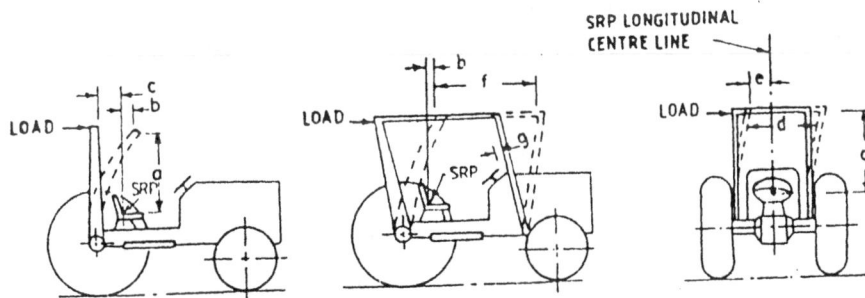

FIG.12.5 ZONE OF CLEARANCE

intrude on the clearance zone described by the dimensions shown in Fig. 12.5. The various dimensions are given below

a= 760 mm at the longitudinal center line

b= not greater than 100 mm to rear edge of crossbar measured from the seat reference points.

c= not greater than 305 mm measured from seat reference point to forward edge of crossbar.

d= minimum 610 mm

e= 50 mm inside of frame upright to vertical centerline of seat

f= minimum 445 mm

g= 50 mm measured from cuter periphery of steering wheel.

viii) Imaginary ground planes

The imaginary ground plane (Fig.12.6) is defined as the surface containing a series of straight lines from the outer edges of the ROPS to any part of the tractor that might come in contact with flat ground and is capable of supporting the tractor in that position if the tractor overturns. For this purpose, the tyres and track width setting shall be assumed to be the smallest standard fitting.

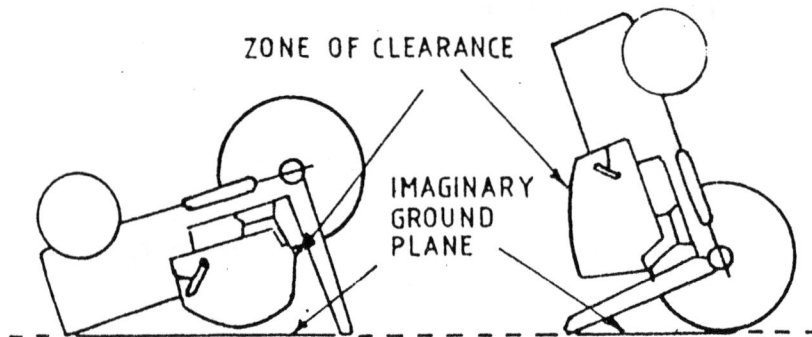

ZONE OF CLEARANCE

IMAGINARY GROUND PLANE

FIG.12.6 IMAGINARY GROUND PLANE

12.8 Test Procedures

A- Checking of specifications and dimensions

The purpose of this test is to check and record the dimensions. mass and specifications of the ROPS and the tractor to which it is fitted. Following checking and measurements should be made :

i) For ROPS

a- Seat reference point, zone of clearance and imaginary ground plane

b- Shape. construction; dimensions. mass and assembling method

c- Mounting method

d- Material used in the construction

e- Accessories if any

f- Label and caution marks

g- Others

ii)For Tractor

a- Dimensions. mass and wheelbase

b- Static sideways overturning angle of the tractor at minimum track width and with ROPS fitted

c- Others

B- Strength Test

The purpose of this test is to determine the strength characteristics of the ROPS and its mountings by simulating such loads as are imposed on the ROPS when the tractor overturns. It consists of

i) - Dynamic strength test
ii) -Static strength test

General condition of strength test

I - Sequence of tests

i) - In the case of dynamic tests
   - Impact at the rear
   - Crushing at the rear
   - Impact at the front
   - Impact at the side
   - Crushing at the front

ii) - In case of static tests:
   - Loading from the rear
   - Crushing at the rear
   - Loading from the front
   - Loading from the side
   - Crushing at the front

The rear loading needs not be performed on ROPS applied to tractors having four wheels drive.

ii) Direction of impacts or loadings

The side chosen for application of side impact or loading on the ROPS shall be that which, in the opinion of Testing Station, can result most unfavorable conditions when series of impacts loading occur. The rear impact shall be applied on the corner opposite the side impact and the front impact on the corner nearest to the side impact or loading.

iii) Removal of components

All detachable windows, doors, panels and non-structural fittings that can easily be removed shall be removed.

iv) Setting of track width

A track width setting chosen for the wheels shall be such that no interference occurs with the ROPS during the tests.

Procedure for dynamic strength test

1) Impact test

The test shall be conducted by applying a dynamic load produced by a pendulum block to the ROPS.

A. Apparatus

i- A pendulum block, which shall be suspended by two chains from pivot points about 6 m above the floor, shall be used.

ii- The pendulum block shall be fitted in such a way that the position of its center of gravity is constant. The mass of the block shall be 2,000+20Kg and its impact face shall have dimensions 680+20mm square.

iii- The lashings used for restraining the tractor to the ground shall be wire ropes of 13 mm diameter.

B. Test conditions

i- Tyre to be used: Standard pneumatic tyres shall be used.

ii-Tyre pressure and deflections: Inflation pressure and deflections in these tyres which are used in the various tests is shown in Table-2

Table 2. Tyre pressure and deflection recommended for different tractors

Type of tractor		Tyre pressure (KPA)bar	Deflection (mm)
Four wheel drive but front and rear wheels are of same size			
	Front	100(1.0)	25
	Rear	100(1.0)	25
Four wheel drive but front wheel are smaller than rear			
	Front	150(1.5)	20
	Rear	100(1.0)	25
Two wheel drive			
	Front	200(2.0)	15
	Rear	100(1.0)	25

C   Test procedure

(a) Impact at the rear

i- Positioning of the tractor
The tractor shall be placed in relation to the pendulum block so that the block will strike the ROPS when the supporting chains and the impact face of the block are at an angle of 20 degree to the vertical. If the angle of the member of ROPS at the point of contact at the moment of maximum deflection is greater than 20 degree to the vertical, the angle of the block shall be further adjusted by an additional support so that the impact face of the block is approximately parallel to the ROPS.

ii-Suspended height of pendulum block
The suspended height of block shall be so adjusted that the locus of its center of gravity passes through the point of contact.

iii- Point of impact
The point of impact shall be that part of the ROPS which would be likely to hit the ground first in a rearward overturning accident, normally the upper edge.

iv-Tractor tie-down
The tractor shall be lashed down by means of steel wire ropes incorporating tensioning devices to ground rails rigidly attached to the concrete base. The points of attachment of the lashings shall be approximately 2m behind the rear axle and 1.5m in front of the front axle. The lashings shall be tightened so that the deflections in the front and rear tyres shall be as indicated in B(ii). After the lashings have been tightened a wooden

block of 150 mm square shall be clamped in front of the rear wheels and tightened.

v- Height of lift of the pendulum block
The height of lift of pendulum block i.e. the vertical height of its center of gravity above the point of impact shall be calculated using the following formula:

(I) For M less than 2.000 kg
   $H = 25 + 0.07M$

(II) For M equal to or greater than 2.000 kg
   $H = 2.165 \times 10^{-3} \, MZ^2$
   or $H = 125 + 0.02 \, M$

Which ever gives the heavier impact

where:

H = Height of lift of the pendulum block(mm)

M = Tractor reference mass (Kg)

Z = Tractor reference wheelbase(mm)

(b) Impact at the front

i-Positioning of tractor
Identical to that described in (a)(i)

ii-Suspended height of the pendulum block
Identical to that described in (a)(ii)

iii-Point of impact
The point of impact shall be that part of the ROPS which would be likely to hit the ground first if the tractor overturns side ways while traveling forward, normally the top of the front corner.

iv-Tractor tie-down
The lashings shall be identical to that specified in (a)(iv) but the wooden block shall be clamped behind the rear wheels.

v-Height of lift of pendulum Block
The height of lift of pendulum block shall be calculated using the following formula

(I)For M less than 2,000 kg
   $H = 25 + 0.07 \, M$

(II)For M equal to or greater than 2,000 Kg

$$H = 125 + 0.02M$$

Where

H= Height of lift of pendulum block(mm)
M= Tractor reference mass(kg)

(c)Impact at the side

i-Positioning of the tractor
The tractor shall be placed in relation to the pendulum block so that the block will strike the ROPS when the supporting chains and the impact face of the block are vertical as shown in Fig. 12.7

ii-Suspended height of pendulum block
Identical to that specified in (a)(ii)

iii-Point of impact
The point of impact shall be that part of the ROPS likely to hit the ground first in a sideways overturning accident, normally the upper edge.

iv-Tractor tie-down
The tractor axle on the side to be struck shall be lashed to the ground rails by means of steel wire ropes. After lashing, a wooden block of 150 mm square shall be clamped against the side of the front and rear wheels opposite the blow and then tightened against them. In addition a beam shall be placed against the rear wheel rim opposite the blow and secured to the floor as shown in Fig.7. The length of this beam shall be so chosen that its position against the rim is at an angle of 30±3 degree to the horizontal.

v- Height of lift of the pendulum block
The height of lift of the pendulum block(Hs) as shown in Fig.7 shall be calculated using the following formula:

(I) For M less than 2,000 Kg

$$Hs+ 25+0.20M$$

(II) For M equal to or greater than 2,000 Kg
$$IIs = 125 + 0.15 M$$

Where

Hs: Height of lift of pendulum block(mm)
M : Tractor reference mass(mm)

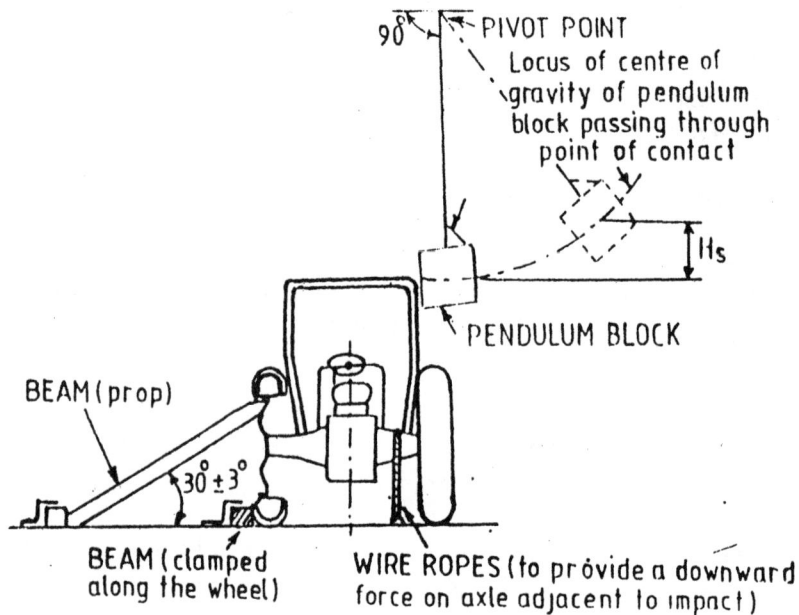

D   Measurements to be made

i-   Defects such as fracture and crack
ii- Permanent deflections of the ROPS
iii- Any part entering in the zone of clearance during test.
iv- Features presenting a serious hazard to the operator
v-   any other point

2)   Crushing test·

The test which is carried out by applying a vertical downwards force on the ROPS through a beam placed laterally across the uppermost members of the ROPS are called crushing test.

A.   Apparatus

Test rig consisting of a rigid beam approximately 250 mm wide, connected to the load-applying mechanism by means of universal joints as shown in Fig.12.8.

B. Test Procedure

a)   Crushing at the rear

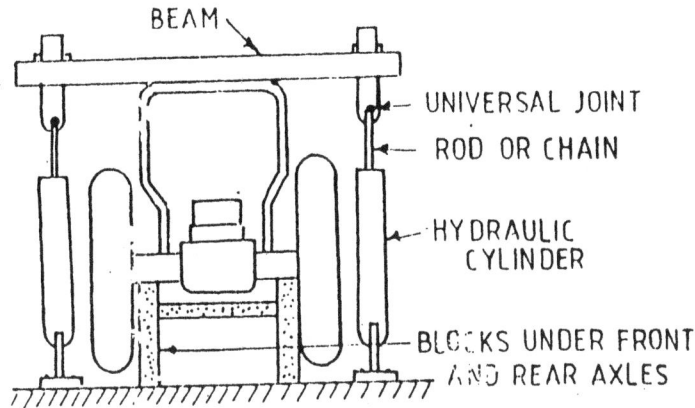

FIG.12.8 METHOD OF CRUSHING TEST

i- The tractor shall be positioned such that the rear edge of the beam is over the rear most top part of the ROPS and the median longitudinal plane of the tractor is midway between the points of application of force to the beam.

ii- Blocks shall be placed under the axles of the tractor so that the tyres do not bear the crushing force.

iii-The crushing force to be applied shall be calculated using the following formula

$$F = 20 M$$

Where

F: Crushing force(N)
M: Tractor reference mass (Kg)

iv- The force shall be maintained for approximately 10 seconds after reach of full force as calculated above.

v- Where the roof of the ROPS is not designed to sustain the full crushing force, the force shall be applied untill the roof is deflected to coincide with theplane joining the upper part of the ROPS with that part of the tractor capable of supporting the tractor's mass when overturned.

(b) Crushing at the front

The procedure is same as followed for crushing at the rear.

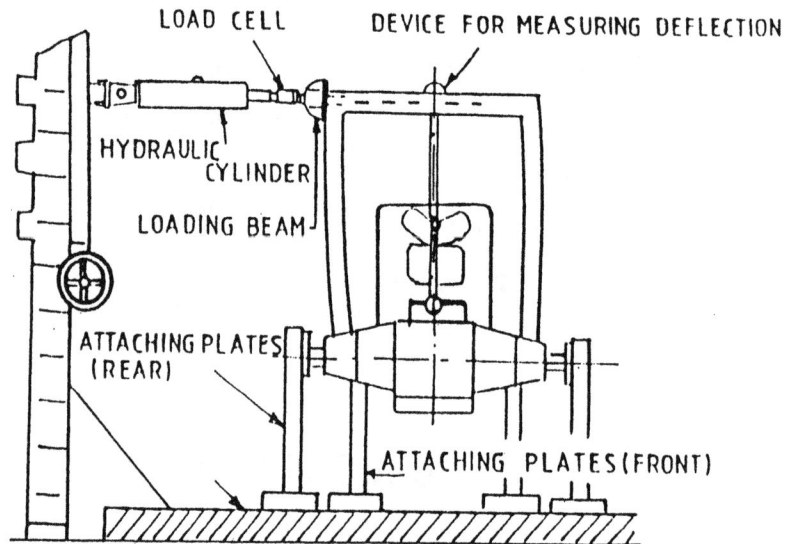

FIG.12.9 METHOD OF HORIZONTAL LOADING TEST

3) Measurements to be made

All items as indicated for impact test.

Procedure for static strength test

Horizontal loading tests
These tests shall be conducted by applying a horizontal static load to the ROPS on test rig as shown in Fig 12.9.

A Apparatus

i) The static testing rig comprises of anchoring rails, supporting plates and hydraulic loading device with load distribution beam.

ii) The load distribution beam used for applying a horizontal force to the ROPS shall have a vertical face having dimension of 150 mm and a length of one the multiples of 50 mm, between 250 mm and 700 mm.

iii) A read out device for measuring force and deflection in order to compute energy absorbed by the ROPS as shown in Fig. 12.10.

B Test conditions
i) The tractor chassis shall be fixed firmly to the ground rails by means of plates,

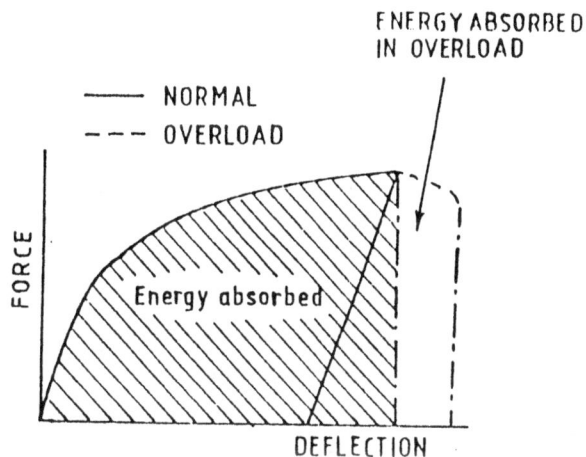

FIG.12.10 FORCE DEFLECTION CURVE

independent of the tyres, to prevent movement during the tests.

ii) The loads should be applied to the ROPS by means of beam. If necessary, a substitute test beam which does not add strength to the ROPS may be utilized.

iii)The direction of the loading at start of test, shall be less than ± 2 degree to the horizontal.

iv) The rate of load application(deflection rate)shall be less than 5 mm/sec

v) The loading shall be stopped when the strain energy absorbed by the ROPS is equal to or greater than the required input energy specified in each test.

C Test Procedure

i)Point of load application
The point of application of load shall be that part of the ROPS which would likely to hit the ground first in a rearward overturning accident. normally the upper edge.

ii)Length of beam
The length of load distribution beam shall not be less than one third of the width of the top of ROPS and not more than 49 mm greater than this.

iii) Required input energy
The energy input to be absorbed by the ROPS shall be calculated using the following formula :

$$E = 1.4 \, M$$

or, $\quad ER = 0.143 \, M$

E = Energy input to be absorbed during rear loading(J)
Er = Energy input to be absorbed during rear loading (Kgf-M)
M = tractor reference mass(Kg)

b)Loading from the front

i) Point of load application
The point of application of load shall be that part of ROPS likely to hit the ground first if the tractor overturned sideways while traveling forward, normally the upper edge.

ii) Length of beam
Same as in the case of loading from rear side

iii) Required Input Energy
The energy input to be absorbed by the ROPs shall be calculated using the following formula

$$E = 500 + 0.5 \, M$$
or, $\quad Ef = 51 + 0.051 \, M$

Where
$\qquad$ E = Energy input to be absorbed during front loading (J)
$\qquad$ Ef = Energy input to be absorbed during front loading (kgf-M)
$\qquad$ M = Tractor reference mass (kg)

c)Loading from the side

i-Point of load application
The point of application of load shall be that part of the ROPS likely to hit the ground first in a sideways overturning accidents, normally the upper edge.

ii-Length of beam
The load distribution beam shall be as long as practicable subject to maximum of 700 mm

iii-Required energy inputs
The energy input to be absorbed by the ROPS shall be calculated using following formula:
$$E = 1.75 \, M$$
$$Es = 0.178 \, M$$
Where,

E = Energy input to be absorbed during side loading(J)
Es= Energy input to be absorbed during side loading (kgf-M)
M = Tractor reference mass(Kg)

D) Measurements to be made

i-Fracture and crack observation
ii-Permanent deflection of ROPS
iii-Any part entering the zone of clearance
iv-Features responsible for serious hazard to operator.
v-Any other relevant point

2) Overload test

If any crack is observed during test which cannot be considered as negligible then loading equal to 120% of the original required energy. shall be applied after the loading test and observation made.

3) Crushing test

This test is similar to dynamic strength test.

Final inspection

The ROPS shall be dismantled and inspected for cracks, fractures and defects etc. after completion of all tests.

12.9 Criteria for acceptance of rops

i-There shall be no feature presenting a serious hazard to the operator during the tests.

ii-There shall be no serious defects affecting the operation of the tractor fitted with ROPS.

iii-There shall be no serious fracture or cracks in all major structural members mounting components or tractor parts contributing to the strength of the ROPS during the tests.

iv- No part of the ROPS shall enter the zone of clearance and come in contact with the seat during the test.

h)Hitches and drawbars

Machines used for towings or which are towed should be provided with adequate towing device which is fitted and secured properly and are safe.

i)Guards for hot parts and fire protection

a-Hot parts which the person may touch and cause burns should be guarded.

b-Hot parts such as exhaust pipe should be designed and fitted in such a manner that these do not cause any hazard to person.

c-Spark arrester may be used with exhaust silencer for arresting glowing carbon particle.

j)Guards for divider
The tips of divider being sharp and dangerous should be guarded.

k)Protection of thrown objects
Working parts of machine which may produce thrown objects, such as stones or fragments of crops or cutter knives, should be guarded adequately in such a manner to ensure the safety of the operator in normal operating position.

l)Safety Signs
Safety signs should be attached adjacent to the following parts :

i-dangerous parts which are difficult to protect by safety guard.

ii-Other parts which are required to alert the person from danger

m)Operational easiness

i- User's manuals should be prepared and supplied with every machine.

ii-All machines should be easy to operate.

Noise measurements

Exposing the ear to loud sounds shifts a person's hearing threshold level upward and he can hear only the louder sounds. This frequent exposure will eventually result in a permanent threshold shift and ultimately to hearing loss.

Measuring and evaluation of noise

Sound is usually measured in decibels. Values range from zero to 140 as shown in Table-3. sound level meter usually has three measuring modes, which are "A", "B"and "F" scales. The "A" scale responds to sound more or less the same as the human ear, and the values are given in units of dB(A).

Table 3: Decible level of common sounds emitted from different sources.

Decible level	Common sounds from different sources
0	Acute threshold of hearing
15	Average threshold of hearing
20	Whisper
30	Leaves rustling or very soft music
40	Average residence
60	Normal speech, background music
70	Noisy office
80	Heavy traffic or window air-conditioner
85	Inside acoustically insulated protective tractor cab
90	Standard limit and hearing damage starts when excessively exposure to noise above 90 db
100	Noisy tractor, power mover, motorcycle
120	Thunder clamp, amplified rockmusic
140	Threshold of pain shot gun, near jet taking off.

Permissible hours per day that a person can safety be exposed to sound levels are given in Table 4..

Table-4: Permissible noise exposure

Hours per day that one can safely be exposed to these sound levels

Duration per day (Hours)	Sound level(Decibles)
8	90
6	92
4	95
3	97
2	100
1.5	102
1	105
0.5	110
0.25 or less	115

12.10 Safety precautions

General

i- Although tractor is designed with adequate safety provisions, there is no real substitute for caution and attention in preventing the mishaps. Once an accidents has occurred, it is too late to think about what one should have done.

ii- Remember that tractor has been designed exclusively for agricultural use. Any other application must first be authorized by the manufacturer.

iii- Do not attempt to increase maximum engine speed by tempering injection pump governor.

iv- Do not alter relief valve setting of hydraulic systems (Power steering, hydraulic lift, remote control valves, etc.)

v- Do not operate tractor, if you feel unwell suspend work rather than taking a risk.

vi- Always use steps and grab handles when getting in or out of the cabin.

vii- As far as possible never work without roll-over protection frame or incor rectly fitted on the tractor. Check that fasteners are not loose and that fasteners frame is not damaged in any way. Do not alter cab by welding or drilling.

viii- Read manual thoroughly before attempting to start, operate, service or refuel the tractor. A few minutes reading will save time and trouble later.

Tractor starting

i- Prior to starting the engine. check that parking brake is on. gear and PTO levers are in neutral position.

ii- Make sure that all implements are fully lowered before starting.

iii - Ensure that all guards and protective devices are correctly installed before starting the tractor.

iv -Do not attempt to start the tractor unless sitting on the operator's seat.

v- Tractor should be operated only by responsible person suitably trained and duly authorized.

vi- Keep a First-Aid kit handy.

vii- Do not work with loose garments that could get caught in moving parts. Check that all rotating parts connected to the PTO shaft are well shielded.

viii- Do not run engine in a closed building without adequate ventilation as exhaust gases are dangerous.

ix- Ensure that there is no person or obstacle within range before starting the tractor.

Tractor operation

i- Select the track width most suitable to the work in hand, keeping the tractor stability in mind.

ii- Engage clutch pedal gradually. Abrupt engagement, particularly on uphill or down hill can cause tractor to pitch dangerously.

iii- Disengage clutch for a moment when front wheels start rising.

iv- Respect the highway code during on-road journey.

v- Do not put foot on brake and clutch pedals.

vi- Latch brake pedals during on-road driving, otherwise dangerous skidding may occur when braking.

vii-Do not drive downhill in neutral or with clutch disengaged.

viii- When the tractor is moving, operator should always be sitting on the operator's seat.

ix- Do not get on or off a moving tractor.

x- Always depress the clutch pedal gently.

xi- Do not take sharp turn at high speed.

xii- Safety gauges should be checked time to time.

Towing

i- Adjust the towing attachment correctly to maintain tractor stability.

ii- Drive slowly when towing heavy trailers or wheeled implements.

iii- Preferably trailers should not be towed unless equipped with an independent braking system

iv- Always use drawbar when towing heavy loads. Do not pull from 3-point hitchlinks as tractor could pitch.

v- When towing, do not take turn with the differential lock in, otherwise you may not be able to steer the tractor.

vi- Always operate the tractor at a safe speed. Reduce speed on slopes and curves to prevent roll-over.

vii-When working on sloping ground, do not drive too fast, particularly when taking turn.

Using agricultural implements and machinery

i- Always use matching implements or machinery rated with the tractor horse power. Never use machinery designed for more powerful tractors.

ii-When hitching coupling, never stand between tractor and implement.

iii-Never operate a PTO driven implement without first ensuring that no. one is on or too near the machine.

iv- Check that all rotating parts connecting to the PTO shaft are well shielded.

Stopping

i- Never leave equipment in the raised position while the tractor is stationary.

ii- Ensure that hydraulic system is not under pressure before disconnecting lines.

iii- Hydraulic oil escaping under pressure could cause serious personal injury. Thus when tracing of oil leaks, wear protective shields, glasses and gloves.

iv- Return gear lever to neutral, disengage PTO, apply parking brakes, stop the engine and engage a gear before leaving the tractor seat. Always remove starter switch key before leaving tractor unattended.

v- Park tractor on level ground as far as possible, engage a gear and apply parking brake.

vi- Before attempting to inspect, clean, adjust, repair of service the tractor or attach implement, ensure, that the engine is stopped, transmission is in neutral position, brakes are applied, PTO is disengaged and all moving parts are stationary.

vii- Work on tyres should be carried out only be experienced personnel using proper equipments. Otherwise tyre fittings could cause a serious accident.

Maintenance

i- If engine just finished the work then allow engine to cool down before removing radiator cap. Slowly turn cap to release pressure before removing cap completely.

ii- Disconnect the battery ground lead before starting any work on the electrical system.

iii- Do not fill tank completely when tractor is to be operated in strong sunlight as fuel could expand and escape. Any escaping of fuel should be wiped off immediately.

iv- Tractor fuel may be dangerous. Therefore, never refuel with tractor in motion, near an open flame or when smoking. Preferably it should be filled in the opening.

v- Always keep a fire extinguisher within reach.

APPENDIX - 1

IMPORTANT BIS TEST CODES FOR AGRICULTURAL MACHINERY

1. 5994-1987   Test Code for Agricultural tractor
2. 11442-1985   Operator's field of vision, method of test
3. 11806-1986   Seat reference point, method for determination
4. 19743-1983   Centre of gravity, method for determination
5. 11821-1986   Protective structures, method of test
6. 11859-1986   Determination of turning and clearnace diameter, method for.
7. 12036-1987   Power take-off belt-pulley performance, method of test.
8. 12061-1987   Breaking performance test method.
9. 12180-1987   Noise measurement, method for.
10. 12224-1987   Hydraulic power and lifting capacity method of test.
11. 12226-1987   Drawbar performance, method of test Format for reporting test results.
12. 9164-1979   Estimating cost of farm machinery operation, guide for
13. 9253-1987   Field performance and haulage test, guidelines for
14. 10740-1983   Operating requirement for PTO driven implements, recommenda tions for
15. 12239-1988   Safety and comfort of operator, guide
16. 9935-1981   Test code for power tiller
17. 9980-1988   Field performance evaluation guideline for
18. 6690-1981   Blade for rotavator
19. 10233-1982   Disc plough, tractor operated
20. 6288-1971   Mouldboard ploughs, test code
21. 6638-1972   Cultivator, spring loaded tractor mounted.
22. 6635-1972   Disc harrow trctor. operated
23. 7640-1975   Disc harrow, test code for
24. 4366-1985   Disc tillage
25. 3369-2985   Puddler, animal drawn
26. 11531-1985   Puddler. test code for
27. 9217-1979   Disc, agricultural, test code for
28. 3293-1965   Levelling KARAHA (KENI)
29. 9813-1981   Terracer tractor mounted
30. 3360-1965   Soil scoop
31. 3292-1965   Hoe, hand, three tined, fixed type
32. 10683-1983   Khurpi
33. 7927-1975   Weeder, paddy, manuallyoperated method of field testing for
34. 3372-1965   Bund former
35. 12334-1988   Bund former, tractor operated
36. 2565-1979   Ridger
37. 10254-1982   Share for ridger
38. 3301-1965   Trampler, green manure, animal drawn
39. 6813-1973   Trampler, green manure, animal drawn

40. 6316-1971   Seed cum fertilizer drill test code
41. 9856-1981   Potato planter. test code for
42. 11893-1986   Potato Planter. semi-automatic
43. 11271-1985   Groundnut planter
44. 11976-1986   Sugarcane Planter
45. 6595-1980   Pump, horizontal centrifugal for clear cold fresh water for agriculture
46. 9137-1978   Pump, centrifugal, mixed flow and axial.
47. 9079-1979   Pump, monoset
48. 10805-1986   Foot-valve
49. 11501-1986   Engine monoset pump
50. 3906
(Part-1)1982   Sprayer, compression knapsack non presseure retaining

(Part-II)1982   Pressure retaining
51. 1971-1982 .   Pump, stirrup, sprayer
52. 3652-1982   Sprayer, foot
53. 11313-1985   Sprayer, power, hydraulic
54. 3062-1982S   prayer rocker
55. 2870-1977   Pump, charge, pressure retaining sprayer
56. 3897-1978   Sprayer, atomizer type
57. 1970   Sprayer, compression knapsack
(Part-I)1982   Non-pressure retaining

(Part-II)1982   Pressure retaining
58. 7593-1986   Spraper-cum-duster, pneumatic power operated Knapsack type
59. 5135   Duster-rotary hand
(Part-I)1974   Belly-mounted

(Part-II)1977   Shoulder mounted
60. 10134-1982   Sprayers, manually operated test code
61. 10093-1982   Cut-off device, method of test for
62. 12482-1988   Dusters, manually operated, test code
63. 6025-1982   Knife sections for grain harvesting machines
64. 8122   Combine harvester-thresher.test code for
(Part-1)76   Terminology

(Part-2)76   Performance test
65. 6284-1985   Thresher, power, cereals, test code
66. 11691-1986   Thresher, spike tooth type
67. 9020-1979   Thresher, power safety requirement
68. 9129-1979   Thresher, power,safe feeding devices
69. 12161-1987   Animal cart, test code
70. 7897-1975   Chaff-cutter, test code for
71. 11459-1985   Chaff-cutter, power operated

APPENDIX - II

List or important testing instruments/equipment and test gadgets

S.No. 1	Name of equipments 2	Range 3	Source of availablity 4
A. Power measurements			
1.	Eddy current dynamometer with digital load and speed indicator	10-HP at 1000 rpm 70 HP at 6000 rpm	a) M/s Associated Electrodyne Industries P Ltd.,Pune - 411 004
2.	A.C.Absorption SCR Electrical dynamometer	At 2000 rpm=1.5hp at 4000 rpm =5.4 HP at 6000rpm =8 hp	b)M/s Birdman Chameng Pvt. Ltd. 7-Hastings St. Calcutta c)M/s Industrial & Agril.Engg.Co. Pvt.Ltd. 43, Forbes St. Fort Bombay d)M/s Mak-Elek Engg.Co., 10, Thigalara Periyanna Lane, Silver Jubilee Park Road, Bangalore e)M/s Technololab Instruments Co., 10-A, Industrial Town, Rajaji Nagar, Bangalore f)M/s Industrial Electronic Pvt. Ltd., B-107, Industrial Estate, Rajaji Nagar, Bangalore g)M/s Industrial Process Automation Pvt.Ltd., A-173 Ist Stage, Peenya Industrial Estate, Bangalore h)M/s A-1 Industrial Andheri (East W.) Vaswaji Road, Dheeraj Pen Compound, J B Nagar, Bombay.
3.	Schenck Hydraulic dynamometer (D700-IE)	75KW at 1000 rpm 700 KW at 7500 rpm	M/s Carl Schenck AGL. Postfach 4018, D-6100 Darmastadt, Germany.
4.	Hydraulic water brake dynamometer (UI-30)	20hp at 500 rpm to 250 hp at 1000 rpm	-do-

1.	2.	3.	4.
5	Hydraulic schenck dynamometer (D-1100)	10 Kw at 500 rpm at 740 kw at 1000 rpm	M/s Carl Schenck AGL. Postfach 4018, D-6100 Darmastadt (W.Germany)
6.	Loading car for drawbar performance test		M/s NIAE, Silsoe, U.K.
B	PULL FORCE MEASUREMENT		
7.	Hydraulic dynamao-meter (pull type)	0-2000kgf	M/s Industrial & Agril. Engg. Co. Pvt. Ltd. New Delhi
8.	Spring type tension dynamometer	0-2000kgf	M/s Bharat Process & Mech. Engg. Ltd., New Delhi
9.	Tensiometer for instant indication of load and tension	0-250kgf	M/s Birdman Chenning (P) Ltd. Calcutta
10.	Universal type load cell	0-100kg and above	M/s Integrrated Process Auto-mation (P) Ltd. Bangalore
11.	Strain gauge	0-100kg	M/s Associated Scientific and Engg. Works, New Delhi
C.	MATERIAL TESTING		
12.	Universal testing machine	0-40 tonne	a) M/s Moharch Marketing Enter-prises, 73-A, Central Avenue, Nagpur.
13.	Brinnel & Rockwell hardness testing machine	0-100HRC	b) M/ s Blue Star Ltd. Bhandari House 91-Nehru Place, N.Delhi
14.	Vickers hardness testing machine Model HD-10	5-10kgf	c) M/s Fuel Instruments and Engineers Pvt. Ltd. Inchalkaranj Maharashtra
15.	Pendulum impact testing machine	0-240 kgm	d) M/s Kamal Metal Industriies Astodia Road, Ahmedabad
16.	Spring testing machine		e) M/s Claxo Laboratories Dr. Annie Basant road, Bombay
			f) M/s Mehta & Co. 253, Jawahar Nagar, Goregaon West Bombay

210

1.	2.	3.	4.
			g) M/s P K Scientific Glassware Govindpura Industrial Estate BHEL, Bhopal (MP)
			h) M/s Toshniwal Bros. Pvt. Ltd. 3-E/B, Jhandewalan Extension New Delhi
			i) M/s Monarch Marketing Enterprises 73-A, Central Avenue Nagpur
			j) M/s Electronic Corpn. of India Ltd. Indl. Development Area, Cherlapalli Hyderabad.

D. METROLOGICAL INSTRUMENTS

1.	2.	3.	4.
17.	Steel tape	3M, 15M, 30M	a) M/s Hindustan tools and Hardware, Mart, Bombay
18.	Outside Micrometer set	0-25mm,25-50mm 50-75mm,75-100mm 100-125mm 125-150mm 150-175mm 175-200mm	
19.	Inside micrometer set	500mm to 300mm	
20.	Cylinder bore gauge	35-60mm	b) M/s Staford Elect. & Engg. Co. Bombay Fort-I
21.	Height gauge dial	0-600mm	c) M/s Smith Engg. Co. Bombay
22.	Steel micrometer	0-25mm depth 0-600	d) M/s Haresh Tools Corpn. 109 Ardeshir Dadi St. Bombay - 40
23.	U-type throat micrometer	0-200mm	e) M/s Industrial Engg. 59 Nagdevi Street, Bombay.
24.	Point micrometer	0-25mm	f) M/s Bharat Engg. Stores 113, Nagdevi Street, Cross Lane,
25.	Disc micrometer	0-25mm	Bombay.

1	2	3	4
26.	Spline micrometer	0-25mm	g) M/s Jayanti Lal & Bros. 78, Narayan Dhru Street, Bombay
27.	Ball micrometer	0-25mm	h) M/s APOO Tools Traders, 96, Narayan Dhru Street, Bombay
28.	Vernier caliper dial type	150mm, 200mm, 300mm 600mm 1000mm	i) M/s Mikrotech, 38/39 Hadapsar Industrial Estate, Pune
29.	Vernier depth gauge		j) M/s Grown Engg. Co. 91, Narayanran Marg, Nadvi, Bombay
			k) M/s Precision Tools Traders 72/8 Nagdevi Cross lane, Bombay.
			l) M/s Hindustan Tools & Engg. Company, Ratansi Champsi Path Lohar Chawl, Bombay
			m) M/s Bora Brothers, 5, Hamidia road, Bhopal (MP)
			n) M/s Mahasukhal & Co. 93, Narayan Dhru street, Bombay
E.	SOIL TESTING		
30.	Digital PH meter	0-14 PH	a) M/s Benz Sales, 2402, Main market Trinagar, Delhi.
31.	Soil sheer test instrument		b) M/s Geologists Syndicate Pvt. Ltd. 137, Biplabi Rosh Bihari Basu road, Calcutta.
32.	Cone Penetrometer		c) M/s Associate Instruments Mfgs (India) Pvt.Ltd. B-5, Gillandar House, Calcutta
33.	Sieve shaker with set of sieve		
34.	Bulk density apparatus		
35.	Sampling auger		d) M/s Toshniwal Bros. Pvt.Ltd. 198, Jamshedji Tata Road, Church Gate, Bombay

1	2	3	4
36.	Hydrometer		e) M/s Lawrence & Mayo (India) Pvt.Ltd. 274-Dadabhai Neoroji Road. Bombay
			f) M/s Beautex Instruments(India) BC Old Rohtak Road, Kishanganj Delhi.
			g) M/s Monarch Marketing Ent. 73-A Central Avenue, Nagpur.
			h) M/s ESEM Trading Corpn. Road No.16 Bunglow No.5 Punjabi Bagh Extension, New Delhi
			i) M/s Pradeep Trading Co. 25-B, Inderlok Old Rohtak Road, Delhi
	F. ELECTRIC POWER, VOLTAGE & CURRENT MEASUREMENT;		
37.	Insulation tester		a)M/s Instrument Technique P.Ltd. Hyderabad
38.	Phase sequence indicator		b)M/s Digital Instrumention New Delhi-6.
39.	Digital frequency indicator	0-1000Hz	c)M/s Digital Instruments Corpn. New Delhi
40.	Three Phase balance wattmeter	0-5Kw 0-10Kw 0-30Kw 0-120Kw	d)M/s Oriental Sciences Appratus Workshop, Ambala Cantt.
41.	Ampere meter	5, 10, 15, 60 & 100 amperes	e)M/s Brieht Start Elect. Pune-411002
			f)M/s Bharat Electronics & Elec. Bombay - 400023
42.	Voltmeter	0-500 volts	g)M/s Scientific Instrumentation Delhi-6.
43.	Energymeter	0-10000 Kwh	
44.	Dimmerstate	0-500 volts	h)M/s British Physical Labs.India Ltd. Bangalore

1	2	3	4
45.	Potential transformer	0-200 volts	i)M/s Systonics, Naroda, N.Delhi
46.	Oscilloscope (dual trace)	Freq.range 20MHz	j)M/s Electronics Corpn.India Ltd. Hyderabad
47.	Digital frequency meter	1Hz-10MHz	k)M/s Instruments & Chemicals Pvt.Ltd., Ambala
48.	Functional generator	.01 to 01MHz	
49.	Digital multimeter		
50.	Wheat stone bridge		
51	Kelvin bridge		
52.	Automatic voltage stabilizer	0-5KVA	
53.	Power factor meter		
54.	Temperature recorder	20-600°C	
55.	Two, four & six channel recorders		

G. VACCUM & PRESSURE MEASUREMENTS

56.	Hydraulic In-line tester	0-75 gpm 0-300 l/m	a)M/s Voltas Ltd., New Delhi b)M/s Toshniwal Bros.Pvt.Ltd. New Delhi
57.	Deadweight pressure gauge tester	0-70kg/cm²	c)M/s Optomohar Industries Pvt. Ltd. Bombay d)M/s Venus Trading Co. Bombay
58.	Mercury/water		e)M/s Sohan Traders. Bombay
59.	Manometers for air intake and exhaust gas pressure		f)M/s Smith Engg.Co. Bombay
60.	Vaccum gauge tester	0 to 1 kg/cm² 0 to 10 kg/cm²	g)M/s The Andhra Scientific Co. Ltd., Mudhelipattanam

1	2	3	4
61.	Vaccum gauge tester	0-1kg/cm^2	
62.	Bourden tube pressure guage		
63.	Oil Pressure gauge		
64.	Tyre pressure gauge		
65.	Atomospheric Barometer		
66.	Compression pressure gauge		

H. TEMPERATURE MEASUREMENT

67.	Digital temp.indicator & 6 channel selector switch	0-200^{0}C	a)M/s Thermo Lab.Equipment, Bombay
68.	Thermocouple indicator	0-1000^{0}C	
69.	Industrial thermometer	0-120^{0}C	b)M/s Toshniwal Bros.Pvt.Ltd. New Delhi
		0-20^{0}C	
		0-300^{0}C	c)M/s Scientific Glass Works Bombay
70.	Six channel Temp. recorder	0-200^{0}C	

I. SPEED TIME AND DISTANCE MEASUREMENT

71.	Hand tachometer	0-10000rpm	a)M/s Toshniwal Bros.P. Ltd. New Delhi
72.	Strobometer	250rpm to 18000rpm	b)M/s Meonix. Madras
73.	Stop watch	0-15min	c)M/s K C Raj & Co. Chandni Chowk, Delhi
74.	Digital hand tachometer	0-9900rpm	d)M/s Anglo Swiss Watch Co. 6 Binoy Badal Dinesh Bagh, Calcutta
75.	Non-contact type digital tachnometer	0-9900rpm	e)M/s J Boseck & Co.Pvt.Ltd. 16/5 Jawaharlal Nehru road, Calcutta

1	2	3	4
76.	Panel-mounted digital tachometer	1-9999rpm	f)M/s Swiss India Watch Co. 457, Kalbadevi road, Bombay
77.	Digital stop watch	0-15min	g)M/s Audiotronics, New Delhi
78.	Measuring tapes		i)M/s Systech Pvt.Ltd. Bombay.

J. MASS MEASUREMENT

79,	Physical balance and weight box	0-5 kg.	a)M/s Harinder Scientific Works, Ambala Cantt.
80.	Single pan semi-micro projection balance	0-100 gm & 0-1000gm	b)M/s Eastern Scientific Co. Calcutta
81.	"Owalabor" electric pan balance	0-1 kg	c)M/s Mettler Instruments Switzerland.
82.	Semi indicating balance	0-1 kg & 0-10 kg.	d)M/s India Maching Co.Ltd. Howarh
83.	Electronic precision balance	0-1 kg	e)M/s Crown Engg.Co. Bombay
			f)M/s Avery India Ltd. 8-Narottom Neroji Marg. Ballard Estate Bombay
84.	Platform balance	0-2000kg	h)M/s Y M Pathare & Co. Post Box No.1428, Bombay
			i)M/s Lawernce & Mayo (India)Pvt. Ltd. 274-D Naoroji Road Bombay
85.	Top pan balance	0-10 kg	j)M/s Saples Scale Mfg.Co. P.Ltd. Prospect House. Reghunath Dadaji Street, Bombay
86.	Avery platform balance	300 kg sens.50gm	k)M/s Venkateshware Weighing Machine Co. 139 Motishaw road, Bombay
87.	Spring balance	50kg, 100kg & 200 kg	l)M/s Bharat Weighing Scales & Engg. Syndicate, 60/1, K. Nudy lane, Howrah

1	2	3	4
88.	Top pan plastic balance	1 kg	m)M/s George Salter India Ltd. Chartered Bank Bldg. Calcutta
89.	Other weighing machines & balances		n)M/s India Machinery Co. Ltd. Dassnagar Howrah

K. MOISTURE MEASUREMENT

90.	a)Vaccum Oven/Electric ovan		a)M/s Toshniwal Bros. Pvt. Ltd. New Delhi
	b)Grain moisture tester		b)M/s OSAW, Ambala City
91.	Infrared Moisture balance	0-100%	c)M/s Bharat Heaters, 19 Tardeo Bridge, Diana Cinema Lane, Bombay
			d)M/s Chetan Electric Industries 125, Allied Industrial Estate, Mahim (West) Bombay
92.	Speedy moisture tester for grain/straw and soil	0-20%	e)M/s Temp Industrial Corpn. 1,Lamington Chambers, 394 Bhadkamkar Marg, Bombay
			f)M/s The Scientific Instruments Co.Pvt.Ltd., New Delhi g)M/s Kailash Bros(P) Ltd. Delhi-6

M. FUEL CONSUMPTION MEASUREMENT

93.	Fuel measuring appara- tus(Digital fuel flowtron)	0-20 Kg/h 0.1gm/min	a)M/s Flow-Tron Indo.Parsippany (USA)
94.	Volumetric fuel consumption meter		b)M/s NIAE, Silsoe (U.K.)

N. AIR VELOCITY

95.	Anemometer for meas- urement of air velocity	0-15m/s	M/s Central Scientific Syndicate, Bombay
96.	Air gas flow meter		M/s Toshniwal Bros.Pvt. Ltd. Madras

1	2	3	4

O. RELATIVE HUMIDITY

97.	Psychrometer		M/s Toshniwal Bros. Pvt. Ltd. New Delhi
98.	Hygrometer		-do-
99.	Hygrotherograph		-do-

P. FLOW, CRACK & THICKNESS DETECTION

100.	Ultrasonic flow-detector	1MHz to 10MHz 50mm to 1000mm	M/s Arkel Enterprises, New Delhi
101.	Magnetic crack detector		M/s Industrial Application, Baroda
102.	Ultrasonic thickness guage		M/s Artek Enterprises, New Delhi

Q. CHEMICAL ANALYSIS

| 103. | Carbon & Sulphur apparatus | | M/s Toshniwal Bros (P) Ltd., New Delhi |
| 104. | Muffal furnance | | M/s Wilson Scientific Works, Delhi - 6 |

R. NOISE LEVEL AND VIBRATION TEST

105.	Precision integrating sound level meter	2"microphose 24 to 130dB 20 KHz to 20 KHs	M/s Bruel & Kjaer, Danmark -do-
106.	Octave filter set	10Hz to 20KHz	-do-
107.	Microphone preamp-lifier	20Hz to 20KHz	-do-
108.	Vibration meter	0.3Hz to 15KHz	-do-
109.	High Sens. Accelerometer	Charge sens. 2 -1+2% pc/ms voltage sens. 2 Bmu/MJ	-do-

1	2	3	4
		5%,0.2-9100Hz 10%,.1-12600Hz	
110.	Tunable hand pass filter	2Hz to 20 KHz	M/s Bruel & Kjeer, Danmark
111.	Level recorder	1Hz to 20KHz	-do-
112.	Triaxial accelero-meter	0.2 to 8700Hz	-do-
113.	Portable tape recorder	Tape speed 33.1mm/Sec	-do-
114.	High resolution signal analyser	Freq.range 10Hz to 20KHz	-do-
115.	X.Y. recorder	X-axis o.02 to 500mv/ mm X-axis 0.02 to 1000mv/mm	-do-
116.	High C. Exiciter head	Rated force 112N Freq.range 5Hz to 10KHz	-do-
117.	Power supply	Load output 7.5 D.C. Open circuit output 12V,DC	-do-

SMOKE DENSITY

| 118. | Bosch smoke meter | | a)M/s Tobert Bosch GEMH, Gsottgart, West Germany. |
| | | | b)M/s Mitutoyo Corporation,Tokyo, Japan |

T. ERGONOMICS MEASUREMENT

119.	Bicycle Ergometer		
120.	Expirograph		
121.	Ergograph		
122.	Stethoscope		
123.	Biomedical Telemetry		

1	2	3	4
124.	Anthropo-metic instrument		
125.	Thermophygrograph		
126.	respirtion gasmeter		
127.	Treadmill		
128.	Hand Grip Strength tester		
129.	Audiometer		
130.	Gas Analyzer		

U. MISCELLANEOUS INSTRUMENTS

1	2	3	4
131.	Water level indicator with sensor		M/s Mitutoyo Corpn. Tokyo-108, Japan.
132.	Sunction & delivery head indicator with sensor		-do-
133.	Revolution indicator with sensor		-do-
134.	Torque indicator with sensor		-do-
135.	Personal computer and printer a)Personal computer(PC/XT) Memory(RAM)-680KB Monochrome display color display Graphics display,keyboard floppy disks b)Dot-Matrik printers with graphics devices		M/s Wipro Information & Technology, New Delhi
136.	Electronic calculator		-do-

APPENDIX - III

Proforma for summary of test results of tractor

A) Specifications
Name & model of tractor:
Engine:
   -Bore/stroke(mm)
   -Number of cylinders
   -Cylinder capacity(cc)
   -Rated speed(rpm)
   -Cooling system
Transmission:
   -Number of speeds
   a) forward
   b) reverse
   -Range of speed(kmph)
   -Clutch
   a) forward
   b) reverse
   -Type of PTO
   -Standard pto speed (rpm)
   -Provision for differential lock
Type of Hydraulic system:

Maxium radius of turning space(m)

Construction:
   -Fuel tank capacity (1)
   -Type of brakes
   -Tyre size: front:
           :rear:
   -Track setting(mm)
    front:
    rear:
   -Wheel base (mm)
   -Ground clearance(mm)
   -Ovarall dimensions:
   a) length
   b) width
   c) height
   -Unballasted mass(kg)
   a)front
   b)rear
   c)total

-Ballasted mass(kg)
a) front
b) rear
c) total
height of Trailor Hitch
Recommendations for wet land operation

Recommended matching implement:
-M.B.Plough (no.of bottoms/size(mm)
-Disc Plough (no.of bottoms/size(mm)
-Harrow (no.of discs/spacing (mm)
-Cultivator (no.of tynes/spacing (mm)

B) Performance Date
P.T.O. Performance:
   -Maximum power (kW/Ps)
   -Engine speed corresponding to maximum power(rpm)
   -SFC at maximum power (g/kWh)
   -Power at standard P.T.O. speed(kW)
   -Maximum torque (N.m.)
   -Torque back-up (%)

Drawbar performance, Unballasted/Blassted:
   -Maximum power (kW)
   -Travel speed at maximum power(kmph)
   -Maximum pull (kN)

Hydraulic performance:
   -Maximum power (kW)
   -Corresponding flow rate (1/min)
   -Lift capacity at standard frame (kN)

Brake performance, Unballasted/Blassted:
   -Minimum stopping distance(m)   2
   -Force required to achieve a deceleration of 2.5/Sec (N)
   -Parking performance

Haulage performance:
   -Gross mass, tonnes (type of trailor)
   -Average travel speed (kmph)
   -Distance travelled per litre of fuel consumption (km)

Field performance:
   a) Suitability for wet land operation
   b) Ploughing:
   - Average depth of cut (cm)
   - Area covered (ha/h)

-Fuel consumption (l/h)

C. Cultivation:
   -Average depth of cut (cm)
   -Area covered (ha/h)
   -Fuel consumption (l/h)

D. Harrowing:
   -Average depth of cut (cm)
   -Area covered (ha/h)
   -Fuel consumption (l/h)

Noise level dB (A):
   -Ambient
   -Operator's ear lever

Machinical vibration (Microns):
   -Foot rest
   -Steering
   -Seat etc.

Appendix - IV

Proforma for Chemical Analysis and Hardness test of different components.

Sr. No.	Name of Implement	Name of Component	Hardness (HB/HRC)	Material of Construction									Nearest grade of steel to which conforms
				C	Si	Mn	P	Sm	Or	Mo	Ni	Any other (Specify)	

APPENDIX - V (a)

Data sheet for field testing of tillage implements (Ploughing/harvesting/cultivating etc.)

Date		Name of supervisor		
Operator	Tractor	Type of soil	Time of test from	to	Gear used
Place of test	No.	Implement	Soil moisture % by wt.	

S. No.	Time for 20M (sec)		Wheel Revolution for 20M	Furrow			Time taken at head-land for 1/2 hour (sec)	Temperature °C				
	R.H.	L.H.	(cm) O-rows	Depth (cm)	Width (cm)			Hrs of day	Amb.	Fuel Engine	Transmission	Coolent Ats pr.

Total

Avg.

Stoppage		Remarks
Time (min/sec.)	Cause	

APPENDIX - V (b)

Proforma for summary of field test for tillage implements

Place of test _____ Type of implement _____ Gear used _____

Test team _____ Type of prime-mover _____ Ballast condition ____

Test items	No. of tests									
	1	2	3	4	5	6	7	8	9	10

Date of test

Atmospheric conditions

Relative humidity (%)
Ambient temp. (^{0}C)
Duration of test (h)
Furrow length (m)
Engine speed (rpm)
- No load
- On load

Field conditions :
Previous treatment
Surface treatment

Av. soil moisture (%)
Av. wheel slip (%)
- Front
- Rear
Av. depth of cut (cm)
Area covered (ha/h)
Time required for one ha (h)
Fuel consumption
- l/h
- l/ha
Av. field efficiency (%)
Av. soil inversion (%)
Av. soil pulverization (%)
Av. draft requirement (kgf)

APPENDIX - VI (a)

Data sheet for field testing of seed-cum-fertilizer drill

Date ____ | Operator ____ | Type of seed ____ | Seed moisture ____ | Place of test ____ | Labour requirement ____

Variety of seed ____ | Germination rate of seed before sowing (%) ____ | Time of start ____ | Name of supervisor ____

No load revolutions ____ | Type of fertilizer ____ | Time of end ____

S.No.	Time for 20 m (sec)	Width of sowing for 3 rows (m)	Width of sowing in (20m)	Wheel revolution in (20m) — LH drive wheel	Wheel revolution in (20m) — RH free wheel or supporting wheel	Depth of sowing — Seed (cm)	Depth of sowing — Fertilizer (cm)	Vertical spacing between seed to fertilizer (cm)	Horizontal spacing between seed to fertilizer (cm)	Seed to seed distance in row (cm)	Row spacing (cm)	Seed rate (Kg/ha) — Wt. of seed in (20m)	Wt. of fertilizer in (20m)	Fertilizer rate Kg/ha	Draft (kg)
	1	2	3	4	5	6	7	8	9	10	11	12	13	14	15
1															
2															

APPENDIX VI (b)

Performa for field test results of seed cum fertilizer drill

Test team .... Type of primemover .... Type of seed .... Seed moisture (%)

Place of test .... Labour requirement .... Variety of seed .... Germination rate of seed before sowing (%) .... Soil moisture (%)

S.No.	Date	Time for 20m (sec)	Duration of test	Width of sowing for 3 rows (m)	Wheel revolution in (20m)		Slip (%)	Depth of sowing		Vertical spacing between seed to fertilizer (cm)
					LH Drive wheel	RH Freewheel or Supporting wheel		Seed (cm)	Fertilizer (cm)	
1	2	3	4	5	6	7	8	9	10	11

Horizontal spacing between seed to fertilizer (cm)	Seed to seed distance in row (cm)	Row spacing (cm)	Seed rate (kg/ha) Wt. of seed in (10m)	Wt. of fertilizer in 10 (m)	Fertilizer rate kg/ha	Draft (kg)	Area covered (ha/h)	Fuel consumed (l/h) (l/ha)	Atmospheric condition Ambient temp. (°C)
12	13	14	15	16	17	18	19	20	21

APPENDIX - VI (C)

Data sheet for field testing of Rice Transplanter

Date _____ Operator _____ Variety of Rice _____ Time of Test: From _____

Name of Supervisor _____ Labour requirement _____ No. of leaves per plant _____ To _____

Name of Manufacturer _____ Length of nursery (m) _____ Place of Test _____ Type of nursery _____

| Sl. No. | Time for 20m Revolution for 20m (sec.) | | Wheel Revolution for 20m | Transplanting dimensions | | | | | Time taken at headland for 1/2 hour (Sec) | | Temperature °C | | | | | Atm. Press-ure (mm of Hg) |
	R.H.	L.H.		Depth of 5 rows (cm)	Width between rows (mm)	Spacing between plants (mm)	No. of plants per hill	No. of plants per sq. meter	Hrs. of day	Ambient	Fuel oil	Eng. Coolant	Trans-mission	

Total						

Average _____

Stoppages	Cause	Remarks

Time (min) _____

- Continued --

Appendix VI C (Continued)

Summary for test results of rice transplanter

1. Duration of test (h)
2. Total time stopped (min)
3   Net duraation of test (h)
4. Avg. forward speed (kmph)
5. Slip : a - RH (%)
         b - LH (%)
6. Area covered
7. Fuel consumption
8. Draft requirement (kgf)
9. Avg. depth of transplanting (mm)
10. Avg. width of transplanting (cm)
11. Avg. spacing between hills (mm)
12. Avg. no. of plants per sq.meter
13. Plant missing (%)
14. Avg. no. of plants per hill
15. Rate of work a - ha/h
                  b - h/ha
16. Avg. length of nursery (m)'
17. Avg. no. of leaves per plant

APPENDIX - VII

Area (hectare) irrigated in 8 hours, for different pumping capacities and irrigation depths

Litres/ sec.	Hectares irrigated in 8 hours (depth in cms.)						Hectare centimeters
	2.5 cm	5.0 cm	7.5 cm	10.0 cm	12.5 cm	15.0 cm	
2	0.2	0.1	0.07	0.05	0.04	0.033	0.05
4	0.4	0.2	0.13	0.10	0.08	0.07	1.00
6	0.6	0.3	0.20	0.15	0.12	0.10	1.50
8	0.8	0.4	0.27	0.20	0.16	0.13	2.00
10	1.0	0.5	0.33	0.25	0.20	0.17	2.50
12	1.2	0.6	0.40	0.30	0.24	0.20	3.00
16	1.6	0.8	0.53	0.40	0.32	0.27	4.00
20	2.0	1.0	0.67	0.50	0.40	0.33	5.00
24	2.4	1.2	0.80	0.60	0.48	0.40	6.00
32	3.2	1.6	1.07	0.80	0.64	0.53	8.00
40	4.0	2.0	1.33	1.00	0.80	0.67	10.00
60	6.0	3.0	2.00	1.50	1.20	1.00	15.00
80	8.0	4.0	2.67	2.00	1.60	1.33	20.00
100	10.0	5.0	3.33	2.50	2.00	1.67	25.00
120	12.0	6.0	4.00	3.00	2.40	2.00	30.00

Note: The hectare-centimeters will have to be calculated from the tabulated data and the requirement at site. A suitable pumping set must then be selected from selection charts of centrifugal pump catalogue for desired capacity and head.

Source: Anonymous, 1978

APPENDIX-VIII

Recommended Size of Suction and Delivery Pipe at Various Flow Rate

Flow rate (lit/sec)	Nominal size of suction pipe (mm)	Nominal size of delivery pipe (mm)
0.50	20	20
1.00	30	25
1.25	40	32
1.60	40	40
2.00	40	40
2.50	50	40
3.20	65	50
4.00	65	50
5.00	65	65
8.00	80	80
10.00	100	80
12.50	100	100
16.00	125	100
20.00	125	125
25.00	150	125
30	150	150
40	200	150
50	200	200
60	250	200
80	250	250
100	300	300
125	350	350

Note : Values have been calculated on the basis of flow velocity of 1.5 m/s and 2.0 m/s in the suction and delivery pipes respectively.

Source : Taneja, 1986

APPENDIX - IX

Length of straight pipe in metres giving equivalent resistance of flow in pipe fittings valve etc.

Size (mm)	Standard Elbow	Medium Elbow	Long Radius Elbow	45° Elbow	TEE	Sluice Valve full open	Globe Valve full open	Angle Valve open	Foot Valve or reflex valve
25	0.82	0.70	0.52	0.40	1.77	0.18	8.24	4.57	2.04
40	0.31	1.10	0.85	0.61	2.74	0.29	13.40	6.71	3.05
50	1.67	1.40	1.07	0.76	3.35	0.37	17.40	8.54	3.96
65	1.98	1.65	1.28	0.92	4.26	0.42	20.10	10.00	5.18
80	2.47	2.00	1.55	1.15	5.18	0.52	25.90	12.80	6.10
100	3.35	2.77	2.13	1.53	6.71	0.70	33.50	17.70	8.23
125	4.26	3.66	2.78	1.86	8.24	0.88	42.60	21.20	10.00
150	4.87	4.26	3.35	2.35	10.00	1.07	48.70	25.30	12.20
200	6.40	5.48	4.26	3.05	13.10	1.37	67.10	33.50	16.20
250	7.62	6.71	5.18	3.95	17.10	1.74	88.50	42.60	20.40
300	9.75	7.92	6.10	4.57	20.10	2.04	100.50	51.80	24.40

Source : Anonymous, 1978

APPENDIX-X (a)

Friction head loss in C.I. and M.S. water pipes (Size of pipes in milimeters)

(Head loss in mtrs. per 10 mtrs. length of pipe)

Water flow rate in LPM	25	40	50	65	80	100	125	150	200	250	300
40	1.25										
80	4.69	0.57									
100	7.00	0.809									
150		1.85	0.662								
200		3.13	1.099								
250		4.75	1.67	0.54							
300		6.80	2.26	0.78							
400			3.98	1.31	0.54						
500			5.97	2.00	0.837						
600				2.84	1.163	0.292					
800				5.00	1.983	0.476					
1000					3.017	0.721					
1400						1.393	0.463	0.182			
1800						2.206	0.74	0.305			
2000						2.677	0.887	0.364			
2400							1.25	0.517	0.138		
2800							1.673	0.647	0.165		
3000							1.903	0.753	0.197	0.066	
3500							2.55	1.165	0.262	0.089	
4000							3.35	1.333	0.328	0.113	0.046
4500								1.660	0.415	0.142	0.055
5000								2.050	0.505	0.174	0.067
5500								2.50	0.582	0.200	0.082
6000									0.707	0.233	0.095
7000									0.937	0.312	0.125
8000									0.187	0.407	0.164
9000									0.487	0.497	0.205
10000										0.607	0.250
11000										0.695	0.295
12000										0.807	0.345

Source : Annonymous, 1978

APPENDIX - X (b)

Data sheet for C.F. Pump Testing

Date _____

Name of Supervisor _____

Name of manufacturer _____ | Pump type

Name of _____ | Suction size

Nature of test _____ | Delivery size

Place of test _____ | (rpm)

	Motor make	Capacity	Specified speed (rpm)
	Motor rating	Head measured by	Head (M)
	Phase	Suction lift	Power (KW)
	Motor speed	Suction head measured	Efficiency (%)
	by	by	Discharge (l/sec)
	Power measured	Power measured	Atmospheric pressure
	by	by	Temp. of test liquid

Sl. No.	Speed of pump (rpm)	Suction head (m)	Delivery head (m)	Manometer distance (m)	Velocity head suction(m)	Velocity head delivery(m)	Total head 'H' (m)	Flow			
								Hook gauge initial reading (mm)	Hook gauge final reading (mm)	Head over notch (mm)	'C' Discharge (l/sec)
1	2	3	4	5	6	7	8	9	10	11	12
1											
2											
3											

- Continued -

Appendix X(b) continued

Power					Pump input (kw)	Water (hp)	Pump efficiency (%)	Remarks
Voltage (v)	Current (a)	P.P .meter reading (kwh)	Energy input (kw)	Motor efficiency (%)	Motor input x Motor efficiency			
14	15	16	17	18	19	20	21	22	23

APPENDIX - XI

Norms of important parameters for different types of sprayers as per BIS test codes.

Name of test	Specified requirements as per BIS for different type of sprayers				
	Hand operated continuous knap-sack sprayer	Rocker sprayer	Foot sprayer	Stirrup sprayer	Hand operated compression Knap sack sprayer
	1	2	3	4	5
A- Performance requirement of sprayer					
i) Discharge rate	Mini-500 ml/mt at mini pressure of 200 KPA	1200 ml/mt at pressure of 450 KPA	1200 ml/mt at pressure of 450 KPA	Mini of 400 ml and 500 ml per minutes of 150 KPA and 45 KPA respectively. Min-80%	Pressure development in tank should be min 400 KPA
ii) Volumetric efficiency	Min-80%	Min-80%	Min-80%	Min-80%	
2- Performances requirement of nozzle	i) The nozzle shall provide a rate of discharge per norm. ii) The rate of discharge shall be within ±5% of declared value.	i) The nozzle shall provide a rate of discharge as per norm. ii) The rate of discharge shall be within ±5% of declared value.	i) The nozzle shall provide a rate of discharge as per norm. ii) The rate of discharge shall be within ±5% of declared value	i) The nozzle shall provide a rate of discharge as per norm. ii) The rate of discharge shall be within ±5% of declared value	i) The nozzle shall provide a rate of discharge as per norm. ii) The rate of discharge shall be within ±5% of declared value
B-Constructional Requirements					
3- Piston	Mini height-13mm Mini thickness 2.5mm	Mini height-16mm Mini thickness-3.5mm	Mini height-16mm Mini thickness-3.5mm	Mini height-13mm Mini thickness-2.5mm	Mini height-13mm Mini thickness-2.5mm
4- Pump Cylinder diameter	Not more than 55mm	Not more than 55mm	Not more than 55mm	Not more than 40mm	Not more than 40mm

237

Appendix XI contd.

	1	2	3	4	5
5- Pressure chamber	Min. capacity should be 8 times the piston displacement	Min. capacity should be 8 times the piston displacement	Min. capacity should be 8 times the piston displacement	N.A.	N.A.
6- Strainer	The apertures of the strainer shall be not more than 625/um	The apertures of the strainer shall be not more than 625/um	The apertures of the strainer shall be not more than 625/um	The apertures of the strainer shall be not more than 625/um	The apertures of the strainer shall be not more than 625/um
7- Springs	N.A.	N.A.	Two springs shall be provided. Variations of spring after test shall be not more than ±5%	N.A.	N.A.
8- Lance	The length of lance should be between 500 to 900 mm	The length of lance should be between 500 to 900 mm	The length of lance should be between 500 to 900 mm	The length of lance should be between 500 to 900 mm	The length of lance should be between 500 to 900 mm
9- Cut-off device	Maximum torque required for trigger actuation shall be not more than 3.5 N-m.	Maximum torque required for trigger actuation shall be not more than 3.5 N-m.	Maximum torque required for trigger actuation shall be not more than 3.5 N-m.	Maximum torque required for trigger actuation shall be not more than 3.5 N-m	Maximum torque required for trigger actuation shall be not more than 3.5 N-m
10- Total mass	The mass of sprayer shall be not more than 9.0 Kg.	The mass of sprayer shall be not more than 11.5 Kg.	The mass of sprayer shall be not more than 11.5 Kg.	The mass of sprayer shall be not more than 9.0 Kg.	The mass of sprayer shall be not more than 9.0 Kg.
11- Gasket Test	The gasket shall be deemed to have passed this test, if no leakage observed from points during 8 hrs test	The gasket shall be deemed to have passed this test, if no leakage observed from points during 8 hrs test	The gasket shall be deemed to have passed this test, if no leakage observed from points during 8 hrs test	The gasket shall be deemed to have passed this test, if no leakage observe from points during 8 hrs test.	The gasket shall be deemed to have passed this test, if no leakage observed from points during 8 hrs test.
12- Endurance Test	The sprayer shall be deemed to have passed this test, if no leakage or breakdown is observed during 48 hrs test.	The sprayer shall be deemed to have passed this test, if no leakage or breakdown is observed during 48 hrs test.	The sprayer shall be deemed to have passed this test, if no leakage or breakdown is observed during 48 hrs test.	The sprayer shall be deemed to have passed this test, if no leakage or breakdown is observed during 48 hrs test.	Tank filled with water at 2/3rd of total capacity & at pressure set 400 to 600KPA is required 100 times repitition in order to clear endurance test.

APPENDIX - XII

Norms of important parameters for different types of dusters as per BIS test codes

Name of test	Specified requirements as per BIS for different types of dusters	
	Hand rotary duster (Shoulder mounted type)	Hand rotary duster (Belly mounted type)
1	2	3

A. Performance Requirements

1.	Air output	The fan should be able to deliver not less than 0.8 m³ of air per minute.	The fan should be able to deliver not less than 0.8 m³ of air per minute.
2.	Dust delivery	The delivery rate at max. discharge setting should be not less than 150gm per minute.	The delivery rate at maximum discharge setting should be not less than 150gm per minute.
3.	Dust throw	The duster should be able to throw the dust upto max. distance of one meter.	The duster should be able to throw the dust upto max. distance of one meter.
4.	Hopper capacity	i) The total capacity of hopper should be 0.005 to 0.0075 m³. ii) The tolerance on the declared capacity should be ±5%.	The total capacity of hopper should be 0.004 to 0.006 m³. The tolerance on the declared capacity shoule be ±5%.
5.	Test of agitation	After completion of test the left out dust powder should not be more than 0.5% mass of total. dust.	After completion of test the left out dust powder should not be more than 0.5% mass of total dust.

B. Constructional requirements

6.	Hopper	On the top of the hopper a filler hole of atleast 130mm diameter should be provided.	On the top of the hopper a filler hole of atleast 130mm diameter should be provided.
7.	Suction pipe	The suction pipe shall have an internal dia of not more than 45mm.	The suction pipe shall have an internal dia of not more than 45mm.
8.	Mass of duster	The total mass of the duster should not exceed 8 kg.	The total mass of the duster should not exceed 8 kg.

APPENDIX - XIII

Data sheet for combine harvester:

Date
Place of test:
Time of start
Supervisor:

Time of end: Operator
Gear used: Combine Model Intensity: Variety Grain moisture %
  Tractor H.P. Crop: 'Length Straw moisture (%)
  Height: of ear Grain/Straw ratio

Sl. No.	Time for 20 m for 3 rows (sec.)	Width of out rows from to (M)	Time taken at corners day (sec.)	Hour of day	Temperature						Observations	(a)Details of sample
					Amb.	Fuel	Eng.	Cool	Trans	Hydro		1.Pre-harvest losses g/m 2
					C	C	C	C	C	C		2.Post-harvest losses g/m 2

Atm pressu Hg
mm

Observations:

1. Engine speed (no load)   rpm
2. Engine speed (on load)   rpm
3. Threshing cylinder speed (no load)   rpm
4. Threshing cylinder speed (on load)   rpm
5. Blower speed (no load)   rpm sample time(s)
6. Blower speed (on load)   rpm collected
7. Beater speed (no load)   rpm Distance Outlet
8. Beater speed (on load)   kmph Distance Outlet (M) (Kg)
9. forward speed   kmph
10. Area covered   ha
11. Fuel consumed   litres Grain out let (1)
12. Duration of test   hrs (2)
13. Total time stopped   hrs (3)
14. Net time   hrs (4)
15. Average time lost   Straw outlet (Kg)
16. Average width of cut (Kg) Chaff & Busha outlet
   at corners.
17. Fuel consumption   Grain output
   a) per hour   a) kg/ha
   b) per hectare   b) kg/ha
18. Area covered   c) kg/ha
19. Time required for 1 ha   Time required to unload the grain
20. Height of stubble   tank
21. No.of grains/ear   d) No. of persons, required for handling combine.

	Time lost (min/sec)	
	Stoppages	Cause
Total		
Avg.		

APPENDIX - XIV

Analysis sheet of combine testing

Test conducted at:
Time   Date:

Grain-Straw Ratio:
Weight of 1000 grains:
Name of combine:
Manufactured by:
Variety:

Sl. No.	Sample collected			Sample Analysis						The critical rate of work ha/hr at the time of sample A x 0.36/t (1)x0.36
From	Area (sq.m.)	Time (Sec)	Weight (g)	Representative sample (g)	Healthy threshed grain (g)	Broken grain (g)	Total threshed grain (g) (5+6)	Unthreshed grain (g) 4-(7+8)	Rubbish grain (g)	Total sample (g) (7+8+9)
	1	2	3	4	5	6	7	8	9	10	= (2) / 11
1										
2										
3										
A. Grain Outlet				100, 100, 100						
Average										
Calculated Value										
B. Straw								7+8 for B		
C. Chaff & Bhusa								7+8 for C		
Total										

APPENDIX - XV

Result sheet of combine

Name of crop.. Test conducted at .... Date .... Name of machine ....

Test No. ....

Calculated by: .... Manufactured by: ....

Pre-harvest losses (kg/ha) ....

Grain output		Cutter bar losses (kg/ha)	Grain through put		Straw output		Crop through-put
kg/h	kg/ha		kg/ha	kg/h	kg/h	kg/h	t/h
1	2	3	4		5		6

Losses due to combine (percent by mass)

Collectable		Non-collectable							Cleaning efficiency (%)	Threshing efficiency (%)
Unthreshed from grain outlet (%)	Broken (%)	Header (%)	Stress weeker			Sieve				
			Threshed (%)	Unthreshed (%)	Broken (%)	Threshed (%)	Unthreshed (%)	Broken (%)		
7	8	9	10	11	12	13	14	15	16	17

242

APPENDIX - XVI

Data sheet - Thresher testing

Name of thresher :
Date of test :
Place of test :
Observations
recorded by supervisor :

Name of the crop :
Variety :
Straw-grain ratio :
Moisture content (%)
  a) Grain b) Straw

Speed of drive pulley
  at no load (rpm) :
Speed of driven
Pulley at no load (rpm) :
Cylinder concave clearance (mm) :

Sieve clearance (mm) :
Air output (cum/sec.) :

Sl. No.	Time				Thresh-ing cylinder speed (rpm)	Power consumption		No. of samples	Samples of various outlet(s)							Total qty. of grain mixture collected at main grain outlet(s) (kg)	Total qty. of grain mixture at sieve under flow (kg)
	Start (hrs)	Finish (hrs)	Stop-page oper-ation if any time (h)	Net quan-tity of crop fed (rpm)		Energy meter, reading at start of test (kwh)	Energy meter reading at end of test (kwh)	Power consu-mption for test duration (kwh)		Main grain outlet(s)		Sieve overflow outlet(s)		Straw busha main grain outlet(s)			
										Qty. (kg)	Time (Sec.)	Qty. (kg)	Time (sec)	Qty. (kg)	Time (sec)		

APPENDIX - XVII

Summary of test results for thresher

Name of Thresher_____   Name of supervisor _____

Sl. No.	Item	No. of tests					
		1	2	3	4	5	6
1.	Cylinder speed (rpm)						
	No load						
	On load						
2.	Feed rate (Kg/hr)						
3.	Grain/straw ratio						
4.	Moisture content(%)						
	i)Grain						
	ii)Straw						
5.	Power Consumption (kwh)						
	i) At no load						
	ii) At load						
6.	Broken grain(%)						
7.	Unthreshed grains(%)						
8.	Blower losses(%)						
9.	Spilled grain(%)						
10.	Threshing efficiency(%)						
11.	Cleaning efficiency(%)						
12.	Optimum input capacity(Kg/h)						
13.	Output capacity(Kg/h)						
14.	Corrected output capacity (Kg/h)						

APPENDIX - XVIII

International System of units (SI)

Sl.No.	Physical quantity	Unit	Symbol
1.	Length	metre	m
		millimetre	mm
		centimetre	cm
		kilometre	km
2.	Mass	Kilogram	kg
		gram	g
3.	Time	second	s
		minute	min
		hour	h
4.	Electric Current	ampere	A
5.	Area	square centrimetre	cm^2
		hectare	ha
		square metre	m^2
6.	Force	kilogram force	kgf
		newton	N
7.	Power	metric horsepower	Ps
		kilowatt	kW
8.	Pressure	kilogram force per centimetre square	kgf/cm^2
		pascal	Pa
		millibar	m-bar
9.	Speed/velocity	meter per second	m/s
		kilometre per hour	km/h
10.	Volume	cubic centimetre	cm^3 or cc
		millilitre	ml
		litre	l
11.	Sound level	decibel(A)	dB(A)
12.	Fuel consumption	litres per hour	l/h
13.	Specific fuel consumption	grams per kilowatt-hour	g/kw.h

APPENDIX - XIX

Important conversions

| 1. | Force: | 1 kgf | = 9.80665 N |
| | | | = 2.20462 lbf |

2.	Power	1 hp	= 1.01387 Metric hp (Ps)
			= 745.7 W
		1 ps	= 735.5 W

3.	Pressure	1 psi	= 6.895 Kpa
		1 kgf/cm²	= 98.067 kpa
			= 735.56 mm of Hg
		1 bar	= 100 kpa
			= 10 N/cm²
		1 mm of Hg	= 1.332 m-bar

| 4. | Specific fuel consumption | 1 g/kw.h | = 0.7355 g/hp.h |

5.	Distance	1 meter	= 3.281 feet
		1 feet	= 12 inches
		1 inch	= 25.4mm

| 6. | Torque | 1 kgf.m | = 9.80665 N.m |

7. Torque backup

$$= \frac{\text{Maximum torque x 100}}{\text{Torque at maximum power}} - 100$$

APPENDIX - XX

List of some tractor manufacturers

1.	Eicher	M/s Eicher Goodearth Limited, 59, N. I. T. Faridabad - 121 001 (Haryana)
2.	Escorts	M/s Escorts Limited, 18/4, Mathura Road. Faridabad - 121 007 (Haryana)
3.	Ford	M/s Escorts Tractor Ltd., Plot No.2, Sector-13, Faridabad - 121 001 (Haryana)
4.	Hindustan	M/s Gujarat Tractor Corpn. Ltd., Vishwamitri Rly. Stn., Overbridge, Baroda - 390001 (Gujarat)
5.	HMT	M/s HMT Limited. Tractor Division, Pinjore - 134 101. Distt. Ambala - 134001(Haryana)
6.	Kirloskar	M/s Kirloskar Pneumatics Co. Ltd. Tractor division, Hadapsar Industrial Estate, Pune 411013 (Maharashtra)
7.	Mahindra	M/s Mahindra & Mahindra Ltd., Tractor division, Akurli Road Kandivili East, Bombay 400067(Maharashtra)
8.	Massey Ferguson	M/s Tractors and Farm Equipment Ltd., Hazur Gardens, Sembiam, Madras - 600011 (Tamil Nadu)
9.	Mitsubishi	M/s VST Tillers & Tractors Ltd., Whitefied Road, Mahadevpura P.O., Bangalore - 560048 (Karnataka)
10.	Swaraj	M/s Punjab Tractors Limited, Phase IV, S A S Nagar (Mohali) Ropar - 160055 (Punjab)
11.	Veer Partap	M/s Partap Steels Limited (Automotive Division) 21/3, Mathura Road, Ballabhgarh (Haryana)
12.	Partap	M/s Auto Tractors Ltd., Lucknow-Varanasi Highway, P. Box No.1, Partapgarh - 230001 (U.P.)

BIBLIOGRAPHY

1. Anonymous. OECD, 1970. Standard code for the official Testing of Agricultural Tractor. Organization for Economic Co-operation and Development, PARIS.

2. Anonymous. 1973. Indian Standard. Test Code for Seed-cum-Fertilizer Drill, IS : 6316.

3. Anonymous. 1974. Indian Standard. Specification for hand rotary duster IS : 5135 (P I & II).

4. Anonymous. 1975. Indian Standard Test Code for Stationary Power Thresher for Wheat. IS : 6284.

5. Anonymous. 1976. Mechanical transplanting of paddy. Test Bulletin/Series-2/76 prepared by Tractor Training & Testing Station (Govt. of India) Budni.

6. Anonymous. 1976. Indian Standard. Test Code for combine harvester-thresher, IS : 8122 (Part-I).

7. Anonymous. 1978. Proceedings of the International Agricultural Machinery Workshop held at International Rice Research Institute, Manila, Philippines.

8. Anonymous. 1978. Indian Standard for acceptance test for centrifugal, Mixed flow and axial pumps. IS : 9137.

9. Anonymous. 1978. Kirloskar 100 Years of Service to the Nation. Bulletin by Kirloskar Oil Engine, Pune.

10. Anonymous. 1980. Indian Standard on specification for horizontal centrifugal pumps for clear, cold and fresh water for agricultural purposes. IS : 6595.

11. Anonymous. 1980. Indian Standard Test Code for Agricultural Tractor (Terminology and General guidelines), IS : 5994 (Part-I).

12. Anonymous. 1980. Indian Standard Test Code for Agricultural Tractor (Laboratory and Tract Tests). IS : 5994 (Part-II).

13. Anonymous. 1981. Indian Standard. Test Code for combine harvester Thresher IS : 8122 (Part-II).

14. Anonymous. 1982. Indian Standard. Method of test for manually operated sprayers IS : 10134.

15. Anonymous. 1983. RNAM Test Codes & Procedures for Farm Machinery. United

248

Nations Development Programme, Pasey City, Philippines.

16. Anonymous. 1983. RNAM Test Codes and Procedures for Farm Machinery. United Nation Development Programme Pasey City, Philippines.

17. Anonymous. 1983. RNAM Test Codes and Procedures for Farm Machinery, United Nations Development Programme, Pasey City, Philippines.

18. Anonymous. 1985. Indian Standard. Test Code for Power Thresher for Cereals IS : 6284.

19. Anonymous. 1986. Bureau of Indian Standards Certification Scheme. Procedure for grant of licence.

20. Anonymous. 1986. Perspective for Agricultural Tractor Industry in India. A report prepared by Ministry of Industry, Govt.of India, New Delhi.

21. Anonymous. 1988. OECD Standard Codes for the official testing of agricultural tractors. OECD Publication Service Paris.

22. Anonymous. 1988. Indian Standard Test Code for disc harrows. IS : 7640.

23. Anonymous. 1988. Indian Standard Test Code for Tractor Operated disc plough. IS : 10233.

24. Anonymous. 1990. Indian Standard Test Code for Mould Board ploughs. IS : 6288.

25. Anonymous. 1990. Japanese Industrial Standard on Testing Methods for Centrifugal Pumps, Mixed Flow Pumps and Axial Flow Pumps. JIS-B-8301.

26. Anonymous. 1992. Proceedings of the All-India Seminar on Pumping Systems, Selection, Maintenance and Management held at IIT, Roorkee.

27. Anonymous. 1993. National Conference on Farm Mechanization held at Bhopal.

28. Dass R.S. 1986. Testing of Irrigation Pumps - Test Codes and Procedures and Interpretation of Test Results. Presented kat USAID development and Management Training Project held at CFMT&TI-Budni.

29. Doharey, R.S. and Patil, R.N. Status of Agricultural Machinery Manufacturing in India. Souvenir released during 22nd Annual Convention of ISAE held at CIAE, Bhopal.

30. Garg, I.K. and Sharma, V.K. 1984. Development and Evaluation of a Manually

Operated Paddy Transplanter. J. of Ag. Engg., 21(1 & 2) : 17-24.

31. Garg, I.K. and Sharma, V.K. 1987. Riding Type Engine Operated Paddy Transplanter. Invention Intelligence, 22 (9-10) : 368-374.

32. Gill, G.S. and Mangat, I.S. 1991. Package of Practices in Agricultural Engineering. Directorate of Extension Education, PAU, Ludhiana.

33. Gill, M.S. 1993. Farm Mechanization. Paper presented at National Conference on Farm Mechanization held at Bhopal, March 2,1991.

34. Kalkat, H.S., Sharma, V.K. and Saini, K.S. 1975. Care and Maintenance of Fertilizer Seed Drill. Progressive Farming.

35. Kaul, R.N. and Ramesh Kumar. 1972. Farm Machinery Testing Centre Test Procedures on Thresher, a publication of PAU, Ludhiana.

36. Kepner, R.A., Bainer, Roy and Barger, E.L. 1978. Principles of Farm Machinery. CBS Publishers, Delhi.

37. Liljedahl, J.B., Jurnquist, P.K., Smith, D.W. and Mokoto Hoki. 1989. Tractor and Their Power Units. AVI, New York.

38. Matsuo Yosuke. 1992. Safety Testing of Agricultural Machinery Lecture delivered at BRAIN-JAPAN.

39. Mehta, M.L. 1984. Standardization in the field of Farm Power. Paper presented in RNAM Training Programme in Standardization of Agricultural Machinery, ISI, New Delhi.

40. Mehta, M.L. and Sharma, V.K. 1985. Feasibility of Delinking Cutting and Threshing Systems of a Chaff Cutter Type Thresher. J.of Ag. Engg., 22(2) : 10 - 18.

41. Mehta, M.L. and Sharma, V.K. 1985. Studies on Thrashing System of Chaff Cutter Type Thrasher. J. of Research, PAU, 22(4) : 735-741.

42. Mehta, M.L., Tiwari, R. and Patil, V.A. 1988. Farm Machinery Testing in India. Agril. Engg. Today Vol. 22, No. 5&6.

43. Mehta, M.L., Tiwari, R., Omkar Singh and Patil, V.A. 1988. Studies on Standardization of Seed-cum-Fertilizer Drills. Paper presented during XXVth Annual Convention of ISAE held at Udaipur (Rajasthan).

44. Mehta, M.L., Suneja, Y. and Tiwari, R. 1988. Studies on Testing and Evaluation

250

of a few selected centrifugal pumps. Paper presented during XXVth Annual Convention of ISAE, held at Udaipur.

45. Mehta, M.L., Tiwari, R. and Patil, V.A. 1989. Quality Control in Rural Industries Manufacturing Agricultural Implements 'Udyog Yug'a publication of Haryana Govt.

46. Mehta, M.L. and Patil, V.A. 1989. Role of Testing and Evaluation of Tillage Implements for Minimum Energy Consumption. Paper presented during seminar on Minimum Tillage Technology held at NRFMT&TI, Hissar.

47. Mehta, M.L., Tiwari, R., and Patil, V.A. 1989. Studies on Grain losses by some selected Self Propelled Combines in wheat crop. Paper presented during XXVI Annual Convention of Indian Society of Agricultural Engineers.

48. Mehta, M.L., Tiwari, R. and Patil V.A. 1989. Role of Safety Standards in Farm Machinery. Paper presented at Farm Machinery Safety Day, HAU, Hisar.

49. Mehta, M.L., Tiwari, R. and Patil, V.A. 1989. Role of Safety standards in Farm Machinery. Paper presented during Farm Machinery Safety Day held at H.A.U., Hisar.

50. Mehta. M.L., Tiwari, R. and Patil, V.A. 1990. Testing of Agricultural Machinery. Agricultural Engineering Today, Vol. 14, No. 1 & 2.

51. Mehta, M.L. and Patil, V.A. 1991. Agricultural Machinery Testing in India. Agricultural Mechanization in Asia, Africa and Latin America - Japan Vol. 22, No.3.

52. Mehta, M.M. 1986. Status of Farm Mechanization in India. USAID Development & Management Training Project Programme held at Central Farm Machinery Training & Testing Institute, Budni.

53. Misra, S.K. 1991. Formulation and Implementation of Agricultural Mechanization Strategies in India. Agricultural Mchanization Policies and Strategies in Africa. Commonwealth Secretariat London.

54. Nakra, B.C. and Chaudhary, K.K. 1989. Instrumentation Measurement and Analysis. Tata McGraw-Hill Publishing Company Ltd., New Delhi.

55. Ochiai, Y., Sugiura, Yasuro and Shigeta, K. 1992. Testing and Evaluation of Agricultural Tractor (Riding Type). Lecture delivered at BRAIN-Japan.

56. Pangotra, P.N. 1975. Tractor Testing To-day. Paper presented at 13th Annual

Convention of Indian Society of Agricultural Engineers held at Allahabad Agril Instt., Allahabad.

57. Pangotra, P.N. 1981. Report on Survey on Farm Mechanization in India. Presented for the Asian Productivity Organization at New Delhi.

58. Peter, E.C. 1986. Development of Trainer's skill in selection of power threshers, operational techniques and preventive maintenance. Paper presented during USAID programme held at Central Farm Machinery Training & Testing Instt..Budni (MP).

59. Prasad, J. and Misra, S.K. 1980. Guidelines for testing animal drawn implements. Paper presented at IXth Annual Workshop of Coordinated Scheme for Research and Development of Farm Machinery held at IGFRI, Jhansi.

60. Prasad, J. and Srivastava, N.S.L. 1982. Impact of Agricultural Mechanization on production, productivity, income and employment generation. Agricultural Engineering Today Vol.15 & 16 (1-6).

61. Ram, R.B., Singh, S. and Verma, S.R. 1980. Comparative Performance of Some New and Conventional Tillage Equipment. J. of Ag. Engg., 17(1) : 7-13.

62. Sahay, J. 1986. Elements of Pumps and Tubewells. Agro Book Agency, Patna.

63. Sharma, R.N. 1982. Indian Standards on Testing of Agricultural Machinery. Paper presented at UNIDO/ICAR Mini Workshop on Testing of Agril. Machinery, P.A.U., Ludhiana.

64. Sharma, V.K. and Garg, I.K. 1981. Paddy Transplanters - History of Development and Present Status. A background paper for Inter-Design Workshop Proposed by National Institute of Design, Ahmedabad.

65. Sharma, V.K., Singh, C.P. and Gupta, P.K. 1984. Research on Thrashers at PAU - A Review. Paper presented at All-India Seminar on Role of Agricultural Engineers in Rural Development at New Delhi.

66. Sharma, V.K., Gupta, P.K., Singh, S. and Singh, C.P. 1986. Power Requirements of Different Systems of Spike Tooth Wheat Thrasher. J. of Institution of Engineers, AG Vol. 65, PP. 17-20.

67. Shigeta, Kazuto. 1992. Data Processing and Analyzing Lecture delivered at BRAIN-JAPAN.

68. Singh, C.P., Garg, I.K., Sharma, V.K. and Panesar, B.S. 1982. Design, Development and Evaluation of 10-Row Tractor Mounted Paddy Transplanter. J.

252

of Ag. Engg., 19(3) : 81-89.

69. Srivastava, N.S.L. and Srivastava, P.V. 1993. Projected Demand of Agricultural Machinery for the Year 2000 A.D. Paper published in Souvenir released during 28th Annual Convention of ISAE, held at CIAE, Bhopal.

70. Srivastava, A.C. 1990. Elements of Farm Machinery. Oxford & IBH Publishing Co. Pvt. Ltd; New Delhi.

71. Suneja, Y. Mehta, M.L., Tiwari, R. and Patil, V.A. 1991. Standardization and quality control of centrifugal pumps. AMA, Japan, Vol. 22, No.2.

72. Suneja, Y. Mehta, M.L. Tiwari, R. and Patil, V.A. 1991. Standardization in Centrifugal Pumps - A case study. Journal of the Institution of Engineers - Vol. 72, Part AG-1.

73. Tak-izawa, Nagagoshi. 1992. Testing and Evaluation of Rice Transplanter. Lecture delivered in BRAIN-JAPAN.

74. Takahashi, M. 1992. Testing and Evaluation of Centrifugal Pump. Lecture delivered at BRAIN-OMIYA-JAPAN.

75. Taneja, D.S., Kaushal, M.L., Sondhi, S.K. and Murty, V.V.N. 1986. A manual on centrifugal pumps for irrigation. Department of Soil and Water Engineering, PAU, Ludhiana.

76. Tiwari, R., Mehta, M.L., Singh, A.K., and Patil, V.A. 1989. Studies on standardization of Hand Operated Knapsack Sprayers. Paper presented at XXVIth Annual Convention of ISAE.

77. Tiwari, T.C. 1986. Testing of tillage implements and their evaluation. Paper presented during USAID programme held at CFMT&TI-Budni (MP).

78. Tsuga, Kohnosuke. 1992. Rice Transplanter. Lecture delivered in BRAIN-JAPAN.

79. Verma, S.R., Chauhan, A.M. and Kalkat, H.S. 1977. Multicrop Seed Drill-cum-Planter. Agri.Engineering Today, 1(ii).

80. Verma, S.R., Rawal, G.S. and Bhatia, B.S. 1978. A study on human accidents in wheat thrashers. J. of Ag. Engg., 15(4).

81. Verma, S.R. 1984. Standardization in the field of Harvesting Machinery. Paper presented lat RNAM Training Programme in Standardization of Agricultural Machinery, BIS, New Delhi.

82. Verma, S.R. 1984. Standardization in the field of harvesting and Thrashing Equipment. Proceeding on RNAM Training Programme on Standardization of Agri Machinery. BIS, New Delhi.

83. Verma, S.R. 1985. Problems and Progress in the evaluation and extension of Seed-cum-Fertilizer drills in India. Proceedings of the Silver Jubilee International Conference Organized by IRRI, Los Banos. Phillipines.

84. Verma. S.R. and J. Singh. 1986. Useful Gadgets for Manufafcture of Low-cost Agricultural Machinery. AMA-JAPAN, 17(4) : 58-62.

85. Verma, S.R. 1989. Standardization of Agricultural Machinery in Nigeria - Some Suggestions. Proceeding of first National Colloquim on Standardization of Agril. Machinery held at National Centre for Agricultural Mechanization, Ilorin, Nigeria.

86. Verma, S.R. and Sharma, V.K. 1993. Mechanization in Indian Agriculture, Achievements and Challenges. Changing Scenario of Indian Agriculture. Commonwealth Publishers, New Delhi.

87. Zachariah. P.J. 1976. Farm Equipment Standardization for Agril. Development. Agril Engg. Today Vol. I, No.5 & 6.

www.ingramcontent.com/pod-product-compliance
Lightning Source LLC
Chambersburg PA
CBHW061324190326
41458CB00011B/3888